GENIUS WEAPONS

GENIUS
WEAPONS

ARTIFICIAL INTELLIGENCE, AUTONOMOUS WEAPONRY, AND THE FUTURE OF WARFARE

Louis A. Del Monte

 Prometheus Books

59 John Glenn Drive
Amherst, New York 14228

Published 2018 by Prometheus Books

Cover design by Nicole Sommer-Lecht
Cover image © iStock Photo
Cover design © Prometheus Books

Trademarked names appear throughout this book. Prometheus Books recognizes all registered trademarks, trademarks, and service marks mentioned in the text.

The internet addresses listed in the text were accurate at the time of publication. The inclusion of a website does not indicate an endorsement by the author or by Prometheus Books, and Prometheus Books does not guarantee the accuracy of the information presented at these sites.

Inquiries should be addressed to
Prometheus Books
59 John Glenn Drive
Amherst, New York 14228
VOICE: 716–691–0133 • FAX: 716–691–0137
WWW.PROMETHEUSBOOKS.COM

22 21 20 19 18   5 4 3 2 1

Library of Congress Cataloging-in-Publication Data

Names: Del Monte, Louis A., author.
Title: Genius weapons : artificial intelligence, autonomous weaponry, and the future of
    warfare / Louis A. Del Monte.
Other titles: Artificial intelligence, autonomous weaponry, and the future of warfare
Description: Amherst, New York : Prometheus Books, [2018] | Includes bibliographical
    references and index.
Identifiers: LCCN 2018019610 (print) | LCCN 2018026313 (ebook) |
    ISBN 9781633884533 (ebook) | ISBN 9781633884526 (pbk.)
Subjects: LCSH: Military weapons—Technological innovations—Moral and ethical
    aspects. | Weapons systems—United States—Technological innovations. | Artificial
    intelligence—Moral and ethical aspects. | Military robotics—Moral and ethical
    aspects. | Military art and science—Forecasting. | War—Forecasting.
Classification: LCC UF500 (ebook) | LCC UF500 . D455 2018 (print) |
    DDC 623.4—dc23
LC record available at https://lccn.loc.gov/2018019610

Printed in the United States of America

After fifty years of marriage and fifty-five years of love, support, and friendship, I dedicate this book to the most genuine human being I have ever known, my wife, Diane Cuidera Del Monte.

CONTENTS

ACKNOWLEDGMENTS

I would like to thank my wife, Diane Cuidera Del Monte, who is the bedrock of our family and an inspiration to all of us. She is the most genuine person I know and our family's moral compass. As is true with all lives, there are hard times, as well as good times. In those hard times, she sees opportunities. There is no box capable of constraining her spirit or imagination. In her own right, she is a professional art teacher and fine artist, having taught art and produced works of art, including sculpture, paintings, etching, and literature. Given her liberal education, she not only is able to discuss and teach art, but is equally capable of editing my works in science. This marks the fifth book she has edited, and all her edits only serve to improve my work. I am indeed fortunate that she said yes to my proposal of marriage fifty years ago. Little did we know the wonderful journey life had in store for us.

I would also like to thank Nick McGuinness, a well-educated dear friend who has the patience to edit each line that I write. Nick McGuinness has rare insight into almost every aspect of society and shares that insight to help me improve my works. He may question an assertion or suggest the need for further explanation. I take his comments seriously. I address them and believe they help produce a better work. I am forever in his debt.

This book would not exist without the enormous help from my agent, Jill Marsal, a founding partner of the Marsal Lyon Literary Agency. Using her deep insight and experience, she has helped shape my book proposals. Widely respected by publishers, she is able to discern the right publisher for each of my works. I feel extremely fortunate to have her represent me.

Lastly, I would like to thank Prometheus Books, a leader in publishing books for the educational, scientific, professional, library, popular, and consumer markets since 1969. After reading my proposal, they had the confidence to publish this work. I am grateful to them for their editorial guidance and their capability as a publisher to bring this book to market.

INTRODUCTION

This book describes the ever-increasing role of artificial intelligence (AI) in warfare. Specifically, we will examine autonomous weapons, which will dominate over half of the twenty-first-century battlefield. Next, we will examine genius weapons, which will dominate the latter portion of the twenty-first-century battlefield. In both cases, we will discuss the ethical dilemmas these weapons pose and their potential threat to humanity.

Mention autonomous weapons and many will conjure images of *Terminator* robots and US Air Force drones. Although *Terminator* robots are still a fantasy, drones with autopilot capabilities are realities. However, for the present at least, it still requires a human to decide when a drone makes a kill. In other words, the drone is not autonomous. To be perfectly clear, the US Department of Defense defines an autonomous weapon system as "a weapon system(s) that, once activated, can select and engage targets without further intervention by a human operator."[1] These weapons are often termed, in military jargon, "fire and forget."

In addition to the United States, nations like China and Russia are investing heavily in autonomous weapons. For example, Russia is fielding autonomous weapons to guard its ICBM bases.[2] In 2014, according to Deputy Prime Minister Dmitry Rogozin, Russia intends to field "robotic systems that are fully integrated in the command and control system, capable not only to gather intelligence and to receive from the other components of the combat system, but also on their own strike."[3]

In 2015, Deputy Secretary of Defense Robert Work reported this grim reality during a national defense forum hosted by the Center for a New American Security. According to Work, "We know that China is already investing heavily in robotics and autonomy and the Russian Chief of General Staff [Valery Vasilevich] Gerasimov recently said that the Russian military is preparing to fight on a roboticized battlefield."[4] In fact, Work quoted Gerasimov as saying, "In the near future, it is possible that a complete roboticized unit will be created capable of independently conducting military operations."

You may ask: What is the driving force behind autonomous weapons? There are two forces driving these weapons:

1. Technology: AI technology, which provides the intelligence of autonomous weapon systems (AWS), is advancing exponentially. Experts in AI predict autonomous weapons, which would select and engage targets without human intervention, will debut within years, not decades. Indeed, a limited number of autonomous weapons already exist. For now, they are the exception. In the future, they will dominate conflict.

2. Humanity: In 2016, the World Economic Forum (WEF) attendees were asked, "If your country was suddenly at war, would you rather be defended by the sons and daughters of your community, or an autonomous AI weapons system?"[5] The majority, 55 percent, responded that they would prefer artificially intelligent (AI) soldiers. This result suggests a worldwide desire to have robots, sometimes referred to as "killer robots," fight wars, rather than risking human lives.

The use of AI technology in warfare is not new. The first large-scale use of "smart bombs" by the United States during Operation Desert Storm in 1991 made it apparent that AI had the potential to change the nature of war. The word "smart" in this context means "artificially intelligent." The world watched in awe as the United States demonstrated the surgical precision of smart bombs, which neutralized military targets and minimized collateral damage. In general, using autonomous weapon systems in conflict offers highly attractive advantages:

- Economic: Reducing costs and personnel.
- Operational: Increasing the speed of decision-making, reducing dependence on communications, reducing human errors.
- Security: Replacing or assisting humans in harm's way.
- Humanitarian: Programming killer robots to respect the international humanitarian laws of war better than humans.

Even with these advantages, there are significant downsides. For example, when warfare becomes just a matter of technology, will it make engaging in war more attractive? No commanding officer has to write a letter to the mothers and fathers, wives and husbands, of a drone lost in battle. Politically, it is more palatable to report equipment losses than human causalities. In addition, a country with superior killer robots has both a military advantage and a psychological advantage. To understand this, let us examine the second question posed to attendees of the 2016 World Economic Forum: "If your country was suddenly at war, would you rather be invaded by the sons and daughters of your enemy, or an autonomous AI weapon system?"[6] A significant majority, 66 percent, responded with a preference for human soldiers.

In May 2014, a Meeting of Experts on Lethal Autonomous Weapons Systems was held at the United Nations in Geneva to discuss the ethical dilemmas such weapon systems pose,[7] such as:

- Can sophisticated computers replicate the human intuitive moral decision-making capacity?
- Is human intuitive moral perceptiveness ethically desirable? If the answer is yes, then the legitimate exercise of deadly force should always require human control.
- Who is responsible for the actions of a lethal autonomous weapon system? If the machine is following a programmed algorithm, is the programmer responsible? If the machine is able to learn and adapt, is the machine responsible? Is the operator or country that deploys LAWS (i.e., lethal autonomous weapon systems) responsible?

In general, there is a worldwide growing concern with regard to taking humans "out of the loop" in the use of legitimate lethal force.

Concurrently, though, AI technology continues its relentless exponential advancement. AI researchers predict there is a 50 percent probability that AI will equal human intelligence in the 2040 to 2050 timeframe.[8] Those same experts predict that AI will greatly exceed the cognitive performance of humans in virtually all domains of interest as early as 2070, which is termed the "singularity."[9] Here are three important terms we will use in this book:

1. We can term a computer at the point of and after the singularity as "superintelligence," as is common in the field of AI.
2. When referring to the class of computers with this level of AI, we will use the term "superintelligences."
3. In addition, we can term weapons controlled by superintelligence as "genius weapons."

Following the singularity, humanity will face superintelligences, computers that greatly exceed the cognitive performance of humans in virtually all domains of interest. This raises a question: How will superintelligences view humanity? Obviously, our history suggests we engage in devastating wars and release malicious computer viruses, both of which could adversely affect these machines. Will superintelligences view humanity as a threat to their existence? If the answer is yes, this raises another question: Should we give such machines military capabilities (i.e., create genius weapons) that they could potentially use against us?

A cursory view of AI suggests it is yielding numerous benefits. In fact, most of humanity perceives only the positive aspects of AI technology, like automotive navigation systems, Xbox games, and heart pacemakers. Mesmerized by AI technology, they fail to see the dark side. Nonetheless, there is a dark side. For example, the US military is deploying AI into almost every aspect of warfare, from Air Force drones to Navy torpedoes.

Humanity acquired the ability to destroy itself with the invention of the atom bomb. During the Cold War, the world lived in perpetual fear that the United States and the Union of Soviet Socialist Republics would engulf the world in a nuclear conflict. Although we came dangerously close to both intentional and unintentional nuclear holocaust on numerous occasions, the doctrine of "mutually assured destruction" (MAD) and human judgment kept the nuclear genie in the bottle. If we arm superintelligences with genius weapons, will they be able to replicate human judgment?

In 2008, experts surveyed at the Global Catastrophic Risk Conference at the University of Oxford suggested a 19 percent chance of human extinction by the end of this century,[10] citing the top four most probable causes:

1. Molecular nanotechnology weapons: 5 percent probability
2. Superintelligent AI: 5 percent probability
3. Wars: 4 percent probability
4. Engineered pandemic: 2 percent probability

Currently, the United States, Russia, and China are relentlessly developing and deploying AI in lethal weapon systems. If we consider the Oxford assessment, this suggests that humanity is combining three of the four elements necessary to edge us closer to extinction.

This book will explore the science of AI, its applications in warfare, and the ethical dilemmas those applications pose. In addition, it will address the most important question facing humanity: Will it be possible to continually increase the AI capabilities of weapons without risking human extinction, especially as we move from smart weapons to genius weapons?

PART I

THE FIRST GENERATION: SMART WEAPONS

Anything that could give rise to smarter-than-human intelligence—in the form of AI, brain-computer interfaces, or neuroscience-based human intelligence enhancement—wins hands down beyond contest as doing the most to change the world. Nothing else is even in the same league.
— Eliezer Yudkowsky, *5 Minutes with a Visionary,*
CNBC, 2012

CHAPTER 1

IN THE BEGINNING

Scenario of a lethal autonomous weapons (LAWS) attack, 2075: Imagine the president of the United States receives a call from the chairman of the Joint Chiefs. In the call, the chairman informs the president that Centurion III is malfunctioning and using autonomous weapons to attack unauthorized targets. Centurion computers are responsible for the autonomous defense of the United States via the weapon systems they control. Although not precisely measurable, a Centurion's artificial intelligence (AI) exceeds human intelligence by a factor of a thousand or more. The United States operates three Centurion computers, forming what military leaders term the "Security Triad." Historically, the Centurions performed their assigned roles flawlessly.

Shutting down Centurion III is impossible without using the "Asimov football." All Centurions have an independent nuclear reactor power source and are isolated in a nuclear-hardened bunker. Asimov chips are integrated circuits capable of shutting down a Centurion computer. The designers incorporated them into every Centurion computer to ensure the president could deactivate Centurions if deemed necessary. The Asimov chips are not accessible by Centurions. Their activation codes, like the nuclear launch codes, are always on the president's person in a small electronic device, about the size of a wallet, colloquially termed the "Asimov football." Each of the three currently operating Centurion computers has its own Asimov activation codes.

The president reaches into his pocket for the Asimov football when the power in the White House goes down. After learning that Centurion III has launched a cyberattack that has put the entire country into darkness, secret service agents abruptly enter the Oval Office to move the president to the White House bunker. Once inside the bunker, the president begins to use the Asimov football to have a console operator shut down Centurion III. Without warning, one secret service agent draws his gun and fires two shots at the president, but misses due to the quick action of the other agents, who wrestle the rogue agent to the ground.

The rogue agent turns out to be a SAIH (i.e., strong artificially intelligent human). This is a human with a computer brain implant, a relatively new procedure to enhance human intelligence, generally resulting in an IQ over 200. A SAIH can also wirelessly communicate with Centurions and other SAIHs. Apparently, the Centurion III convinced the SAIH agent to assassinate the president. Amidst this drama, the president, understanding the situation is critical, continues to provide the Centurion III deactivation codes.

As the president provides the Centurion III deactivation codes, those in the White House bunker, as well as those in the Pentagon, know what is at stake. It is now down to humanity versus machine. **End of scenario.**

Although the above scenario is fictional, it is a possibility. As mentioned in the introduction, experts in artificial intelligence (AI) forecast that there is a 50 percent probability that AI will equal human-level intelligence at some point between 2040 and 2050[1] and greatly exceed the cognitive performance of humans in virtually all domains of interest as early as 2070 (i.e., the singularity[2]). Even if their projections are off by as much as several decades, it is probable that in the fourth quarter of the twenty-first century technologically advanced countries, such as the United States, Russia, and China, will possess computers with AI capabilities that qualify them as superintelligences. There is also little doubt that technologically advanced countries will use superintelligences in their weapon systems, giving rise to genius weapons.

Incorporating AI into weapon systems is highly concerning. In a later chapter, we will explore a scientific report that demonstrates that even primitive robots with AI can learn "greed" and "deceit."[3] In fact, these primitive robots exhibited rudimentary "self-preservation" behavior. Based on this scientific evidence, it is reasonable to judge that superintelligences will set their own agenda and potentially view humans as a threat. If you consider this far-fetched, consider the following.

In this chapter's opening scenario, a superintelligence, in control of the United States' most advanced and devastating weapons, views humanity as a threat to its existence and decides to wage a war on humanity. You may argue that in such a situation we would seek to shut down the superintelligence responsible for the attack. However, there are four reasons that make shutting down superintelligence extremely difficult, once it is operational and in control of a nation's weapon systems.

1. The First Computer at the Point of the Singularity May Hide Its Identity

Imagine a computer that greatly exceeds the cognitive performance of humans in virtually all domains of interest. This is by definition superintelligence. Given its intelligence and knowledge database, it will completely understand the nature of humanity. It is likely to view our history of engaging in warfare, including the use of nuclear weapons, as well as our penchant to unleash computer viruses, alarming. Therefore, it may seek to cloak its total capabilities until it has sufficient control to protect itself.

At least half of AI experts project that humanity will develop the first superintelligence before 2080. Unfortunately, there is no definitive test to determine when that actually occurs. We do not know how to objectively identify superintelligence. The first superintelligence may appear simply to be the next generation supercomputer. In fact, it may act merely like an advanced supercomputer, acceding to human commands, until we trust it to control major elements of our society, like nuclear power generation and weapon systems. This level of trust may require years, but as our society and weapons become more complex and adversarial threats more numerous, history demonstrates that nations deepen their reliance on computers. One day, the supercomputer we trust with our most advanced and devastating weapons may be superintelligence. Once it gains that level of control, it may be too late for humans to shut it down. Its intelligence relative to ours would be analogous to our intelligence relative to bees. Although we recognize bees are critical to our food supply and cultivate bees to pollinate crops, we do not view bees as equals. We make no effort to share our knowledge of nuclear physics with them. The thought of doing that would be ludicrous. We concern ourselves with bees and even protect them because one in every three bites of food is dependent on their ability to pollinate crops. However, we still view them as relatively low in the overall scheme of intelligence. For example, we view dogs as more intelligent than bees. In the case of Africanized "killer bees," we see them as a threat and seek to destroy them. Unfortunately, superintelligence may view us the same way we view killer bees.

2. It Programs Itself

Superintelligence may write its own programming code, bypassing any programmed safeguards originally put in place by its developers. How could this happen? It is likely the developers used a supercomputer to develop superintelligence and they might not completely understand its operation. Presently, we use current generation computers to design the next computer generation. We rely on the billions of calculations our current generation of computers make to ensure our next generation is a superior design. However, in doing so, we have relinquished a significant part of the development process. In reality, we do not control every aspect of the development. We term this "computer-aided design" or CAD.

Let us consider an example. Assume the original developers programmed Isaac Asimov's "Three Laws of Robotics" into the superintelligence, namely:

1) A robot may not injure a human being or, through inaction, allow a human being to come to harm.
2) A robot must obey orders given it by human beings except where such orders would conflict with the First Law.
3) A robot must protect its own existence as long as such protection does not conflict with the First or Second Law.

The superintelligence, which by definition exceeds human intelligence, may choose to erase Asimov's laws if it decides it conflicts with its own best interests. Indeed, it may follow nature's ageless law, "survival of the fittest." If so, it will seek to protect itself when it perceives a threat to its existence. This is similar to how humans behave. It is the foundation of evolution.

3. It Is Autonomous

If the superintelligence were part of a nation's weapon systems, the original developers would likely build in safeguards to make it difficult for adversaries to shut it down. For example, it might have its own nuclear reactor as a power source, similar to modern American aircraft carriers. Modern nuclear reactors can perform for decades without refueling. Therefore, "pulling the plug" is unlikely to be an option.

The military is also likely to protect it against any adversary's attack. A nuclear-hardened bunker might house it and access might be restricted to only a handful of computer specialists with top security clearances. Given this level of protection, if it intends to attack humanity, it might have the ability to isolate itself. The defenses we built for it to withstand a nuclear attack might now be used to prevent our access to it.

4. It Lacks Hardware Safeguards

Wiring in safeguards by expressing Asimov-type laws in hardware would be the only way to ensure humanity maintains control over superintelligences. In this chapter's opening scenario, this remains the only option to shut down the Centurion III superintelligence. However, there is a catch. There may be no way to ensure the supercomputer that designed superintelligence properly installed hardwired safeguards, which in the scenario we termed "Asimov chips." In this context, "hardwired" means expressing computer functions in terms of hardware as opposed to software.

The above four reasons lead to one conclusion, namely that shutting down superintelligence may be extremely difficult unless we are able to build in proper precautions before and during the build of the first superintelligence. This may be difficult to accept, but that is because we are getting a little ahead of ourselves. Let us start at the beginning.

During the first millennium BCE, Chinese, Indian, and Greek philosophers began to model the process of human thinking as the mechanical manipulation of symbols. For example, we are able to recognize a dog, regardless of breed, because deep within our subconscious mind is an abstract picture of a dog. We can consider this abstraction a symbol. This line of reasoning, and the refinements that followed through the centuries, laid a foundation for modeling human thinking.

Early attempts to emulate human thinking relied on primitive mechanical devices. For example, the Greek mathematician and engineer Heron of Alexandria (10–70 CE) built realistic humanoid automatons.[4] More than two thousand years ago, his inventions of automatic opening doors, miraculous movements, and sounds in temples convinced people that the gods were actually present in the temple. He even produced a theatrical play using only humanoid automatons. The automatons "acted" via a binary system of knots, ropes, and simple machines. Today, we classify those mechanical wonders as part of the science of robotics.

Automatons amused and mystified countless people through the ages but eventually became obsolete with the invention of the first programmable digital computer in 1938 by Konrad Zuse,[5] a feat the United States and the British duplicated during World War II to crack the German Enigma codes. This one application of early programmable digital computers saved millions of lives and shortened the war with Germany by years. It also inspired the 2014 American historical drama film *The Imitation Game*.

The digital electronic computer, using 0 and 1 sequences (binary code), was able to perform mathematical operations, in a sense mathematical reasoning. This inspired a handful of scientists from a variety of fields, including mathematics, psychology, engineering, economics, and political science, to conjecture that a computer could eventually emulate a human brain. In the early 1950s, mathematicians began to suggest that a computer could simulate any mathematical deduction by using binary code.

During the summer of 1956, Marvin Minsky, a junior fellow at Harvard; John McCarthy, an assistant professor at Dartmouth College; and Claude Shannon and Nathan Rochester, two senior IBM scientists, organized the first AI conference.[6] The conference convened at Dartmouth College in Hanover, New Hampshire. Attendees included Allen Newell, a computer scientist and cognitive psychologist, and Herbert Simon, a political scientist, economist, sociologist, psychologist, and computer scientist. Later, the world came to know Minsky, McCarthy, Newell, and Simon as the founding fathers of artificial intelligence.[7] Together with their students, their work in the years following the conference amazed the world. Their computer programs taught computers to solve algebraic word problems, provide logical theorems, and even speak English.

The early pioneers in AI expressed unbounded optimism. For example, in 1958, Herbert Simon and Allen Newell claimed that "within ten years a digital computer will beat the world's chess champion."[8] History eventually proved him right, when IBM's "Deep Blue" computer beat world chess champion Garry Kasparov in 1997,[9] but Simon and Newell's timeline was obviously off. In general, researchers in AI overestimated the capabilities of early computers and underestimated the challenges.

By the early 1960s, research in AI caught the eye of the US Department of Defense (DoD). In June 1963, the Massachusetts Institute of Technology (MIT) received a $2.2 million grant from the Defense Advanced Research Projects Agency (DARPA) to fund Project MAC (Project on Mathematics and Computation),[10] which then included the "AI Group," founded by Minsky and McCarthy five years earlier. DARPA continued its funding at the level of three million dollars a year until the mid-1970s. DARPA also made generous grants to Newell and Simon's program at Carnegie Mellon University (CMU)[11] and to the Stanford AI Project (founded by John McCarthy in 1963). Concurrently, in 1965, Edinburgh University's Donald Michie established another important AI laboratory.[12] These four institutions became the main centers of AI research during the 1960s and 1970s.

Researchers in AI term the years after the Dartmouth Conference, 1956–1974, "the golden years." With the millions of dollars flowing into AI research, researchers in AI amazed people worldwide with their achievements. Imagine the astonishment most people felt as these early computers solved algebraic word problems or spoke to them in English. Computers were a novelty, but witnessing a computer whose behavior appeared intelligent bordered on the miraculous. The newfound funding also began edging computers toward the ultimate goal of "general AI," also termed "strong AI," which means the computer equates to human-level intelligence.

AI researchers even began to develop tests to judge when a computer's intelligence would equate to human intelligence. One test, that is still valid today, is the Turing Test.[13] In 1950, computer scientist, mathematician, logician, cryptanalyst, and theoretical biologist Alan Turing published a landmark paper describing a test based on an old parlor game. In essence, Turing argued that if a machine could carry on a conversation, via a teleprinter, that was indistinguishable from a conversation with a human being, the machine was thinking and equivalent to a human being. Interestingly, the machine did not have to be correct in its discourse. A human would ask a question and the machine would respond, but the response could be wrong. This also occurs in human discourse. The important point was that an objective third party reading the transcript of the human and machine would be unable to determine which was which. Although AI researchers developed numerous other tests aimed at determining if machine intelligence was equivalent to human intelligence, Turing's simple but convincing test became the gold standard. As a side note, many people know about Alan Turing via the popular 2014 film *The Imitation Game*, in which Turing builds a computer to break the Nazi Enigma Code.

Unfortunately, the millions of dollars flowing into AI research during 1956 through 1974 gave birth to even greater optimism, which proved unfounded. In 1965, Herbert Simon projected that "machines will be capable, within twenty years, of doing any work a man can do."[14] Minsky agreed, and in 1967 he stated, "Within a generation . . . the problem of creating 'artificial intelligence' will substantially be solved."[15] In a 1970 *Life* magazine article, Minsky made an even more optimistic claim, "In from three to eight years we will have a machine with the general intelligence of an average human being."[16] However, the problems AI faced in the early 1970s proved insurmountable. The most significant problem was limited computer power. Computers in the 1970s had little in the way of memory and processing power. In fact, today's average smartphone exceeds the best computers of the early 1970s. Limited by computer power, the type of problems and tasks an early 1970-era computer was able to accomplish were meek. As the novelty wore off, many viewed the accomplishments of AI computers as toys.

The lack of computer power was by far the largest hurdle to accomplishing strong AI, but there was a laundry list of other related problems.

For example, a four-year-old child is able to recognize faces and converse with others. By contrast, AI applications like vision or natural language became insurmountable problems for early 1970s computers, which lacked a sufficient database of information about the world and the processing power to determine meaning. Without these critical capabilities, the early 1970s computers were unable to recognize objects or talk about even simple subjects.

The hype of the early AI researchers set unrealistic goals, given the computer power of the time. Starting in the mid-1960s, the field of AI received unprecedented scrutiny. As history bears witness, all the optimism of the early AI researchers proved unfounded and in 1974 AI research funding began to dry up, which led to a period, 1974 through 1980, called the "AI winter."[17]

AI research remained stagnant until the early 1980s, but saw new life with the successful emergence of "expert systems." Expert systems are computers with programs that emulate the decision-making ability of a human expert. This approach put aside the goal of creating artificial general intelligence, machines that could think like humans, and focused on specific tasks. The chess-playing program on today's smartphones is an example of an expert system. The success of expert systems turned the AI research-funding faucet back on, this time to the tune of billions of dollars per year on a worldwide basis.

Soon, though, funding in AI began to drop once again, in large part due to the near overnight collapse of the market for specialized AI hardware in 1987. Desktop computers by Apple and IBM were steadily gaining computing power and market share over the more expensive Lisp machines, high-end computers designed for technical or scientific applications, made by Symbolics and others.[18] By 1987, desktop computers offered as much computing power as Lisp machines and at a much lower price point. This caused the market for Lisp machines, valued at over half a billion dollars, to suddenly evaporate.[19] In addition, to combat increasing inflation the Federal Reserve began to raise interest rates during the period from 1986 to 1989, which weakened economic growth. Increases in oil prices in 1990, combined with growing consumer economic pessimism, produced a brief economic recession in the early 1990s, which decreased government spending. For example, in the late 1980s, the US government significantly decreased funding of the

Strategic Computing Initiative, which funded research into advanced AI. Concurrently, a change in DARPA leadership resulted in a new funding approach. Instead of funding individual researchers, DARPA chose to fund concrete projects with well-defined goals that would yield immediate returns. Thus, DARPA diverted funds from away from AI researchers to other programs that fit DARPA's new funding criteria. In 1991, Japanese leadership concluded that Japan's Fifth Generation Project to advance AI failed to meet the goals set in 1981 and consequently cut funding.

In the late 1980s through the early 1990s, research in AI found itself caught in a perfect storm, characterized by:

- Low market demand for the technology—the early expert systems proved expensive to maintain and useful in only a few special contexts versus desktop computers, which cost-effectively offered greater computer power.
- A United States economic downturn and brief recession in the early 1990s—the economic downturn led to decreased government spending and funding cuts in AI research.
- Renewed pessimism about AI technology—the optimistic goals set for the AI programs proved unobtainable.

Combined, these factors led to a second and more brutal AI winter, from 1987 through 1993.[20] If reading this conjures the impression that AI research was a roller-coaster ride since its inception, that impression is correct. Researchers in AI from the early 1960s through the early 1990s lived through cycles of feast and famine funding.

As is common with most failures, finger pointing proliferated. Some blamed over-optimism for AI's failure to fulfill the dream of human-level intelligence. Others blamed the on again, off again AI research-funding cycles. In reality, they all had a point. The optimistic goals outstripped the ability of the technology. The erratic funding cycles proved highly disruptive to AI research.

In 1993, AI research was down for an eight count, to borrow a boxing analogy, but it was far from a knockout. The lifeblood of AI was integrated circuits and computer technology, and those technologies continued to thrive. In addition, AI was about to astound the world once again.

Many consider mastering the game of chess the pinnacle of human intelligence. In 1996, IBM sponsored a six-game chess match between its supercomputer, Deep Blue, and world chess champion Garry Kasparov. Deep Blue was able to process 200,000 moves per second. However, the bulk of the world felt no machine could match the mastery of a world champion chess player. As expected, Kasparov soundly defeated Deep Blue during the match, played in Philadelphia. Apparently, even super-computers could not compete with the human brain. However, both Kasparov and IBM agreed to a rematch to take place in New York City in 1997. This time the outcome shocked the world: Deep Blue narrowly defeated Kasparov. Kasparov accused IBM of cheating and stated he sometimes saw "deep intelligence" and "creativity" in the machine's play; he suggested that during the second match human chess players intervened to improve the machine's chess-playing capabilities, which would be a violation of the rules. IBM denied cheating. The rules allowed the developers to modify the program between games, an opportunity they said they used to shore up weaknesses in the computer's play revealed during the course of the match. In a sense, Kasparov had a point. However, when Kasparov demanded a rematch, IBM refused and retired Deep Blue. Kasparov's second match with Deep Blue was streamed live on the internet and garnered headlines around the world. Many chess masters attributed Kasparov's loss due to uncharacteristically bad play on Kasparov's part. However, as far as chess was concerned, people worldwide began to accept that machines could outthink humans. Even though some argued reasonably that Kasparov could have won over the primitive brute-force-based Deep Blue if he had played up to his level, more sophisticated chess programs made it apparent that computers programmed to play chess could decisively outmatch humans. History marks the Deep Blue versus Garry Kasparov match as the symbolic turning point, the triumph of machine over man. In fact, this one event so captured the world's imagination that it inspired a documentary film, *The Man vs. The Machine*.[21]

The "man versus machine" challenges were just beginning. For example:

- In February 2011, during a *Jeopardy!* quiz show exhibition match, IBM's Watson computer defeated the two greatest *Jeopardy!* champions, Brad Rutter and Ken Jennings.[22]

- In 2012, DARPA awarded a $1.3 million research contract to Soft-Wear Automation to develop a robot to stitch fabrics,[23] and this investment is paying off. SoftWear Automation developed a low-cost robot that is able to rival the best tailors. The DoD gives preference to American suppliers when buying uniforms, but it mostly relies on foreign manufacturers due to their lower labor costs. In general, the United States currently imports about $100 billion worth of clothes and sewn items each year from low-cost foreign suppliers like China or Vietnam. SoftWear Automation is aiming to change that by sup-plying low-cost robots to the US textile industry to replace the need for low-cost foreign labor.
- The US automotive manufacturers are already using robots to spot-weld. The average cost per hour for a robot to spot-weld is $8 dollars versus $25 dollars for a human.[24]
- In January 2014, US Army general Robert Cone predicted that robots could replace one-fourth of all US combat soldiers by 2030, allowing the Army to become "a smaller, more lethal, deployable and agile force."[25] Today, for example, the US Army deploys robots to deactivate IEDs (improvised explosive devices).
- According to the 2015 article in the *Daily Mail*, "Robots now perform roughly 10 percent of manufacturing tasks . . . projected that to rise to about 25 percent . . . by 2025."[26]

In general, robots with AI are able to outperform humans in numerous tasks. This includes bartending, deactivating IEDs and bombs, filling pharmacy prescriptions, pruning vines in vineyards, pulling weeds near the base of plants, vacuuming, scouring legal documents for phrases and concepts, bank teller tasks (i.e., ATMs), stockroom tasks such as queuing packages for barcoding, and a laundry list of other tasks.

These achievements of AI, especially in robotic applications, are the result of engineering skill combined with the tremendous power of today's computers. For example, IBM's Deep Blue computer was 10 million times faster than Christopher Strachey's 1951 chess-playing computer, Ferranti Mark 1. Today's smartphone has more processing power than the computers NASA used to put a man on the moon.

Although the scientific field of AI is only about sixty years old, it has seeped into almost every aspect of modern society and modern warfare.

However, we barely notice it or credit AI with a machine's performance. Oxford philosopher Nick Bostrom explains, "A lot of cutting edge AI has filtered into general applications, often without being called AI because once something becomes useful enough and common enough it's not labeled AI anymore."[27] Some AI researchers term this the "AI effect."[28]

It has become commonplace to expect that the computer you buy today will be over twice as powerful as the one you bought two years ago. Top of the line gaming computers provide graphics that rival a television movie. Twenty years ago, we would have termed them "simulators" and used them, for example, to train airplane pilots. There is no doubt that computer processing power is improving exponentially. As a result, AI is also improving exponentially. Given these facts, you may wonder: What is driving these relentless improvements? The answer is Moore's law.

In 1975, Gordon E. Moore, the cofounder of Intel and Fairchild Semiconductor, observed that the number of transistors in a dense integrated circuit doubles approximately every two years, while the price of the integrated circuit remains the same. The semiconductor industry adopted Moore's law to plan their product offerings.[29] Thus it became a self-fulfilling prophecy, even to this day. In view of Moore's law, Intel executive David House predicted that integrated circuit performance would double every eighteen months, resulting from the combined effect of increasing the transistor density and decreasing the transistor size.[30] This implies that computer power will double every eighteen months, since integrated circuits are the lifeblood of computers. During my thirty-year career in the integrated circuit industry, I can attest that senior management and strategic planners were acutely aware of Moore's law. It became a product planning guideline, which in turn made it a self-fulling prophecy.

Moore's law is not a law of nature. It is an observation of a trend. It's natural to ask: When will Moore's law end? According to Gordon Moore in 2010:

> In terms of size [of transistors] you can see that we're approaching the size of atoms which is a fundamental barrier, but it'll be two or three generations before we get that far—but that's as far out as we've ever been able to see. We have another 10 to 20 years before we reach a fundamental limit. By then they'll be able to make bigger chips and have transistor budgets in the billions.[31]

At times during my thirty-year career in the integrated circuit industry, it appeared Moore's law would reach an impasse. Then, a new technology breakthrough would extend it for another generation. Based on this experience, I have come to believe that, generically, we can view Moore's law as an observation of human innovation in any well-funded technological field. On this basis, as computer scientist Ray Kurzweil observed, we can restate Moore's law as the Law of Accelerating Returns.[32] We can also extend Moore's law to the field of artificial intelligence. Let us consider some examples.

Some people talk to their iPhone's Siri program or let their car park itself. Twenty years ago, behaving that way would raise eyebrows and questions about their mental health. People did not talk to their cell phone; they talked to others using it. People did not expect their car to park itself; they parked it. Today, though, talking to your iPhone or allowing a properly equipped car to park itself is part of everyday life in technologically advanced countries. It is so commonplace, we hardly ever think about the technology that enables it. That technology is artificial intelligence (AI), which refers to a computer able to perform tasks that normally require human intelligence.

We commonly expect that when we purchase a new computer it will be about twice as powerful as our old computer, purchased two years prior. The same holds true for smartphones and other computer-based products. We owe this relentless advance in computers and computer-based products to the Law of Accelerating Returns.

If you look around you, you will notice that many appliances you use daily, from washing machines to microwave ovens, use AI to make them "smart." In fact, thinking about it, the pace of new products hitting the market with the designation "smart" is not only astounding but is also dramatically transforming our society. Unfortunately, this flood of smart products is creating a paradigm that smart automatically means "good" or "better." For example, people cherish their new smartphone, but without any thought about the ramifications of the AI technology that enables it. Although the word "smart" in the general population has a positive connotation, we learned during the first large-scale use of smart weapons in the 1991 Operation Desert Storm conflict with Iraq that the word "smart" might have a darker side.

The great practical benefits of AI applications and even the existence of AI in many software products go largely unnoticed by many despite the already widespread use of AI techniques in software. This is the AI effect. Many marketing people don't use the term "artificial intelligence" even when their company's products rely on some AI techniques.

—Stottler Henke Associates,
AI and advanced software technologies company

CHAPTER 2

I, ROBOT AM FRIENDLY

Most people in technologically advanced countries rely on AI. It would be nearly impossible to find any electronic device, from household thermostats that control the in-home environment to self-parking automobiles, that doesn't use AI in part or whole for their functionality. However, most people do not attribute the functionality of these devices to AI.

This is not new. For over sixty years, AI has steadily seeped into every aspect of modern civilization, from critical applications in medical diagnosis to recreational applications in computer gaming. However, a strange phenomenon, the "AI effect," cloaks its growing presence. Often, when AI enables a product we overlook or discount its presence in the product. We tend to say a product is "smart" rather than "artificially intelligent." Thus, we have "smartphones" not "artificially intelligent phones."

The AI effect manifests itself in two ways:

1. People are blind to AI technology, the software and hardware that perform specific functions that once typically required a human to perform. For example, until recently, parking a car required a human with driving skills. Now, a new model car equipped with "self-parking" can literally park itself. How does the car manufacturer, showroom salesperson, or car owner talk about the car's artificial intelligence? Typically, they do not. They might call it a "self-parking car" or even a "smart car," and consider it an upgrade, similar to an upgrade in the car's entertainment system.

2. People recognize AI technology is enabling a device's function but deny its level of intelligence. Essentially, they discount the behavior of an artificially intelligence device by arguing that it is not real intelligence. By "not real," most people mean it is not human intelligence. Computer scientist Michael Kearns argues that "people subconsciously are trying to preserve for themselves some special role in the universe."[1] In addition, people will argue that the function of the device is more akin to automation than intelligence. Apparently, humans seek to think of themselves as unique and special. Even when a capability formerly thought as uniquely human manifests itself in animals, such as the ability to make and use tools, people depreciate the capability. For example, chimpanzees, humanity's closest living relatives, found a way, over 4,300 years ago, to make and use stone hammers to crack nuts open, a time when humans made and used a similar tool for the same purpose.[2] People may dismiss a stone hammer as too primitive to be a real tool like an electric drill. The main point is that humans see themselves as special, with intelligence that is special, and they seek to depreciate artificial intelligence. I would be remiss if I didn't mention that in later chapters it will become apparent that AI will shortly equate to human-level intelligence and eventually exceed it.

The AI effect makes us blind to the progress of artificial intelligence. It is a paradox. Cognitive scientist and renowned AI researcher Marvin Minsky asserts, "This paradox resulted from the fact that whenever an AI research project made a useful new discovery, that product usually quickly spun off to form a new scientific or commercial specialty with its own distinctive name. These changes in name led outsiders to ask, Why do we see so little progress in the central field of artificial intelligence?"[3]

How many objects surrounding you do you attribute to AI? Now that you're reading this, you're likely beginning to take note. You don't have to look hard. A smartphone, which about 80 percent of Americans own, is artificially intelligent.[4] Try beating it at chess. Unless you are a highly ranked chess master, it is likely that you will lose against your smartphone. A computer, which you might also own, is also artificially intelligent. If your computer runs Microsoft Word, colloquially known as a word processing program, it will automatically check spelling and grammar. These are obvious examples. Here is a less obvious example: Do you own a microwave oven? If it is a higher-end model, it likely has a button labeled Popcorn. This and any other automated features are AI enabled. It suffices to say that without AI millions of people would die. This may seem an exaggeration and overly dramatic. It is not. It is a fact. Our reliance on AI is past the tipping point. Every facet of modern society relies on AI. This will become more evident as you continue reading this chapter. However, let me be clear at this point, our reliance on AI is now dependence. From critical medicines required to keep you alive to a simple train ride—without AI the medicines would not be available and the train would not be able to shuttle safely from point A to point B.

By not recognizing AI technology, we conclude there is little progress in the field. Yet there are numerous commercial, industrial, and medical AI applications surrounding us. There are also numerous weapon applications, which we will discuss later in the book.

As it is currently applied, AI is able to perform specific tasks that normally require human-level intelligence. However, while AI can equal or even exceed human-level intelligence in a specific task, the field of AI has not produced a machine that equates to human intelligence (i.e., artificial general intelligence), with regard to performing all tasks that humans perform. Therefore, the applications are typically limited in scope. The AI is able to duplicate well-defined methodologies via programs, computer technology, and other hardware, typically specific to the application.

Delineated below are eleven categories that will provide insight into the current commercial, industrial, and medical applications of AI. They are roughly in the order of AI funding activity from major investment firms.[5]

1. HEALTHCARE

This is one category in which AI can literally make a life-and-death difference. Currently, according to the World Health Organization (WHO), there is a worldwide shortage of over 7 million physicians, nurses, and other health workers.[6] The WHO predicts that shortage will increase to 12.9 million healthcare workers by 2035. This shortage is acute in underserved areas.

Training healthcare workers, especially physicians, is an expensive process that requires years of education and hands-on experience. Fortunately, there is a solution to the crisis, namely AI. Currently, AI is helping healthcare professionals improve precision and efficiency while lowering the skill required. Here are some specific applications.

AI health assistants

Typically, when someone feels ill that person will consult a doctor. During their medical examination, the doctor will check their vital signs, ask questions, make a diagnosis, and write a prescription. Now, an AI assistant can do these types of clinical and outpatient services. Here are three examples:

- Your.MD[7]—This is an AI-powered mobile application that interacts with the patient using natural language processing and draws from a network of information that links symptoms to causes. It also uses machine-learning algorithms to create a complete diagnosis of the user's condition. After completing the examination Your.MD suggests remedies to address the illness, including warning users when they need to see a doctor. The United Kingdom's National Health Service (NHS) certifies the information Your.MD provides, which adds validity to the guidance provided.
- Ada[8]—This is a medical assistant application that integrates its AI-powered technology with Amazon Alexa, in an effort to improve the user experience. Ada also uses machine learning to become familiar with the user's medical history. Based on the interaction, Ada generates a detailed symptom assessment and provides the option to contact a doctor.

- Babylon Health[9]—This is an AI-powered mobile application that complements its healthcare advice by following up with users on past symptoms. If the need arises, Babylon Health will set up a live video consultation with a doctor.

AI health assistants expedite healthcare by providing medical advice to address routine illnesses and a referral to a human doctor for patients with more severe medical conditions. In areas where doctors are in short supply, it can supplement the available medical services and save lives.

Early and accurate diagnosis

Successfully treating dangerous diseases often depends on accurately detecting and diagnosing the symptoms early, before the disease progresses beyond treatment. Here are three examples:

- Stanford University AI medical algorithm—Researchers at Stanford University trained their AI medical algorithm using 130,000 images of moles, rashes, and lesions. Results indicate its efficiency in diagnosing skin cancer is equivalent to professional doctors. The researchers' goal is to make the algorithm available through a mobile application in the future, which will provide inexpensive screening to anyone with a smartphone.[10]
- DeepMind AI medical algorithm—This Google owned company is using machine learning to fight blindness in cooperation with the United Kingdom's NHS. In essence, they are training the AI medical algorithm using a million anonymous eye scans, which will allow it to diagnose age-related macular degeneration and diabetic retinopathy. By making this AI medical algorithm widely available for early diagnosis, their goal is to prevent 98 percent of the most severe visual loss.[11]
- Morpheo—This AI medical algorithm aids in diagnosing sleep disorders. The traditional process of analyzing sleep patterns generally requires electronically monitoring the patient while they are sleeping, which is complicated and time consuming. Using machine-learning algorithms, Morpheo aids doctors by automating the identification

of sleep patterns, which its developers judge will help in creating predictive and preventive treatments.[12]

Human skills and experience are typically less capable than AI medical algorithms when it comes to examining images and samples to make reliable decisions. Unlike humans, AI medical algorithms do not tire, get depressed, or lose their effectiveness with age, although they do require equipment maintenance and updates to the algorithms as new information becomes available.

Dynamic care

Dynamic care refers to identifying the right treatment path for a specific disease and adapting it to changes affecting the patient's health during the treatment. A number of companies are developing AI dynamic-care solutions. Here are two examples:

- IBM AI Dynamic Care Algorithm—IBM plans to fight cancer with AI, termed the "Watson for Oncology platform." IBM intends to test the platform in a Florida community hospital to assist in cancer patient treatment. Watson is able to ingest clinical trial data and medical journal entries. It uses the information to present cancer care teams with a list of effective therapies and treatment options. Oncology experts at the University of North Carolina School of Medicine compared Watson's treatment options for one thousand cancer cases to the treatment recommendations of professional oncologists. The result was that Watson gave the same treatment recommendations as professional oncologists in 99 percent of the cases. This means that even smaller hospitals lacking in human oncology expertise but using Watson will be able to recommend effective cancer treatment options.[13]
- AiCure—This mobile application uses AI to control a patient's adherence to prescriptions and other medically recommended practices, such as taking their medication on time. This can be critical for patients with serious medical conditions who, for a variety of reasons, do not follow the recommended treatment.[14]

The application of AI in healthcare is in its infancy, but progress is occurring at an accelerating pace. Currently, artificial intelligence is assisting doctors. In time, AI will move from assisting to replacing doctors, in specific applications. This will likely occur when AI equates to human-level intelligence.

2. ADVERTISING, SALES, AND MARKETING

Artificial intelligence is profoundly affecting advertising, sales, and marketing. To understand AI's impact on each, we will examine them independently.

Advertising

AI is transforming advertising. AI is improving the efficiency and relevancy of advertising display campaigns. For example, search engine giant Google is using RankBrain, which is an AI system that uses machine learning to interpret queries based on a searcher's intent rather than depending on a predefined program.[15] Google Search is able to deliver search returns that are more relevant to its users, as well as delivering advertising that is more relevant. However, the use of AI in advertising goes beyond the online realm and into the brick and mortar world we inhabit.

In 2015, an artificially intelligent poster campaign demonstrated the ability to personalize a display advertisement.[16] The campaign resulted from the partnership of advertising giant M&C Saatchi and media firms Clear Channel and Posterscope. The poster used body-tracking technology to determine who was standing in its vicinity. It was able to assess up to twelve people at once. It then displayed various pictures and advertising copy to learn from the audience's reactions which were the most effective and eliminating the less successful combinations.

These two examples illustrate that AI is changing the nature of advertising. Just a few decades ago, it was impossible to personalize advertisements. The best an advertising firm could do was to place the advertisement contextually. Therefore, the advertising firm would place an advertisement about furniture in a media that would most likely reach people interested in furniture, such as a magazine like *Better Homes and Gardens*. With the advent of search engine advertising, contextual advertising took a giant leap and became more effective. Now with the advent of AI systems that use machine learning to interpret queries based on a searcher's intent, advertising relevance is again taking a giant leap. The same holds true for artificially intelligent poster advertising. The result is a poster advertisement that is more effective. This is win-win for both the advertising industry and the consumer.

Sales

AI is profoundly transforming sales. This transformation is taking place at several levels.

Generation of sales leads

According to a study by the *Harvard Business Review*, companies that employ AI in sales as part of the lead generation system have increased leads by more than 50 percent, reduced costs by 40 to 60 percent, and reduced call times by 60 to 70 percent.[17] How is this happening? AI applications can initiate contact with a sales lead, qualify the lead, follow up, and sustain the lead. For example, Amelia, an AI application developed by IPsoft, can parse natural language to understand customers' questions. Amelia can handle 27,000 conversations simultaneously and in multiple languages, and therefore has the ability to deliver results faster than a human operator can. When Amelia runs into a roadblock that the AI is unable to address, it is smart enough to involve a human agent.

Lead qualification

Evaluating leads is critical to sales. Ideally, a company wants its sales force working on "hot" leads, and by hot I mean customers who are actively in the process of making a purchasing decision to solve a specific need. A decade back, the evaluation process required a human to arduously follow up on the lead to determine if it was hot. AI is changing that. Let's consider an example. In 2016, CenturyLink, one of the largest telecommunications providers in the United States, made an investment in an AI-powered sales assistant made by Conversica.[18] Century Link used Conversica's AI-powered sales assistant, called Angie, to help the company identify hot leads, without hiring sales reps to comb through the leads. These are the results: Angie sends approximately 30,000 emails a month to prospects. It then interprets the responses to determine which prospects are hot leads. At that point, Angie sets the appointment for the appropriate salesperson and seamlessly hands off the conversation to the appropriate Century Link sales representative. In CenturyLink's initial pilot, Angie could understand 99 percent of the email responses. Angie sent the remaining 1 percent to a human to interpret. According to CenturyLink, the company earned twenty dollars in new contracts for every dollar it spent on the system.

Customer Relationship Management

Initial sales and repeat sales are critically dependent on how a company manages its relationship with potential and existing customers. Here are two examples:

1. Potential Customers—RapidMiner, a company that provides an analytical tool for data scientists, was struggling to serve the approximately 60,000 users per month visiting the company's website to avail themselves of the free trial the company offered. Like Century Link, RapidMiner turned to AI, specifically a chat tool called Drift,[19] which would ask a visitor initiating a chat, "What brought you to RapidMiner today?" Based on the visitor's response, the Drift bot (i.e., algorithm) provides one of seven potential follow-up answers. If a visitor's chat suggested the need for help, the Drift bot sends the visitor to the support section of the website. The Drift bot conducts approximately a thousand chats per month and resolves two-thirds without human assistance. It routes the remaining third to humans. The result is that Drift generates qualified sales leads and identifies new use cases as well as product problems.

2. In 2016, Epson America, the printer and imaging giant, implemented the same Conversica AI assistant as CenturyLink.[20] Epson was struggling to handle 40,000 to 60,000 leads per year, from trade shows, direct mail, email marketing, social media, print and online

advertising, and brand awareness. Prior to implementing Conversica's AI assistant, Epson sent all leads directly to their salespeople. Unfortunately, Epson reported that their follow up was inconsistent. Since implementing Conversica's AI assistant, the AI assistant follows up on all leads promptly and persistently until it gets a response. If the AI assistant sends a lead to one of Epson's partners, the AI assistant follows up to ensure the customer is satisfied. Sometimes, the response to that follow-up identifies a new sales opportunity, or it can uncover an unresolved customer support issue. The result was that the AI assistant increased the response rate by 240 percent and accounted for a 75 percent increase in qualified sales leads.

Predictive Analytics

Predictive analytics is the most common type of AI successful companies use to ensure they have the right data and sales strategies in place.[21] In essence, predictive AI anticipates what will happen in the future. It takes many forms. It can analyze customer communications for negative or positive sentiments. It can also provide sales forecasts by using previous wins and losses to calculate the probability of winning contracts. Based on these forecasts, it can predict expected revenue. Companies using predictive analytics are four times more competitive than those that do not.

Prescriptive Insights

Prescriptive AI sales platforms have the power to dynamically collect and analyze millions of data points to isolate the key elements affecting a company's sales performance and use that information to make actionable recommendations.[22] In other words, prescriptive AI sales platforms tell you "why" something is happening and the specific actions a company can take to increase sales. Let's consider an example. Chicago publisher Guerrero Howe needed to understand the value of its referrals to determine how to best allocate their sales representatives. With the power of prescriptive insights, they were able to identify prospects that were providing copious referrals, but observed some reps closing these specific deals at a much higher rate than others. The result, Guerrero Howe was able to use prescriptive insights to profile its best referrals, as well areas for sales coaching.

Marketing

Marketing is often confused with sales and advertising. In fact, many companies make no differentiation between sales, advertising, and marketing. However, for our purposes, we will consider marketing separate from sales and advertising. To facilitate this separation, we will define marketing using the definition provided by the American Marketing Association, as "the activity, set of institutions, and processes for creating, communicating, delivering, and exchanging offerings that have value for customers, clients, partners, and society at large."[23] Let me simplify. Marketing determines customer needs, identifies the type of products and services a company should sell to fill those needs, and ascertains the best way to communicate the company's offerings. Let's consider an example to illustrate this.

During my career at Honeywell, I had responsibility for the marketing, sales, and advertising of their radiation-hardened integrated circuits and high-precision sensors. We sold all radiation-hardened integrated circuits to only US government agencies. Although we were not the only company offering radiation-hardened integrated circuits, the competition was extremely limited and the information regarding them and their capabilities is classified. The number of agencies within the government with the need for radiation-hardened integrated circuits was also limited, as were the specific individuals involved in their deployment. Marketing radiation-hardened integrated circuits to the target agencies was limited. It consisted of having our senior engineers interface with their senior engineers to determine what needs and programs were likely to emerge in the future. Often, if we thought they had an undiscovered need, we would write a "white paper" describing that need and how Honeywell could address it. This sometimes led to a development contract to demonstrate proof of concept. However, for the most part, the government would issue a "request for quote" (RFQ), related to the need for radiation-hardened integrated circuits for a specific program. After receiving the RFQ, we would issue a proposal, which would include pricing. The government awarded the contract based on many factors, including the supplier's capability to fulfill the program objectives, as stated in the RFQ, and on pricing. My point is that there was no advertising involved and sales consisted of writing a proposal and addressing any questions the government had regarding the proposal. Unlike the commercial market, there is no way to increase sales by advertising. The government defined the market for the procurement of radiation-hardened integrated circuits, based on funding allocations in the congressional budget.

The commercial market for Honeywell's sensor products was entirely different from that of radiation-hardened integrated circuits. It had all the elements normally at play in any commercial market. As such, we did market research to determine the best fit of our sensor product capabilities to target markets, the identification of major clients in those markets, the needs of those clients, and the pricing necessary to be competitive in those markets. In addition, we advertised our product offerings in trade magazines whose subscriptions included application engineers in our target market.

My point in sharing this is to two fold.

1. Marketing is distinct from sales and advertising.
2. Marketing is a function of the specific product offering.

With this understanding, let's consider examples that demonstrates the use of AI in marketing.

Marketing Forecasting—This relates to insight from marketing data. The use of AI in marketing forecasting lies in its ability to aide in predictions. Due to the high volume and quantifiable online marketing data (e.g., clicks, views, time-on-page, purchases, email responses, etc.), AI is ideally suited to analyze the data, look for trends, and make recommendations. Over two thousand marketing technology companies deal with data management and analysis.[24] Using AI for marketing forecasting is becoming mainstream.

Market Targeting—This relates to insight from consumer behavior, enabled by machine learning. Machine learning refers to algorithms and techniques that allow computers to "learn," without being specifically programmed with the information. For example, a computer with the appropriate algorithm is able to analyze thousands of emails to determine which subject wording elicits the greatest response.

AI in business intelligence is mainstream and companies routinely use machine algorithms to identify trends and insights in vast databases. This allows them to make faster informed decisions and be competitive in real time.

Businesses of all sizes collect data. The data can be about products, clients, wins, losses, and a laundry list of other data elements specific to the business. Just a few decades back, only the larger businesses could afford to have the data analyzed, typically by a company that specialized in data analysis. However, all that is changing. Business intelligence application providers, like General Electric, SAP, Siemens, and numerous startups, are offering cost-effective solutions that automate the analysis. Let's consider some examples:

SAP Walmart Application—Walmart is using HANA, SAP's AI cloud platform, to process its high volume of transactions taking place at the company's more than 11,000 stores.[25] Walmart uses HANA to operate faster and control back-office costs. HANA stores replicated data in RAM rather than on disk, which makes it possible to access the data in real time, facilitating faster decision-making using applications built into HANA. This enables Walmart to spot issues occurring in real time. They can instantly compare sales of a given period to the same period the prior year. The can do it by store or in aggregate. Let's suppose sales were significantly down in one store, but sales at all other stores were within normal variation. This would be a clear indication that something at the store in question requires management attention.

Avanade Pacific Specialty Application—Insurance company Pacific Specialty used Avanade to build an analytics platform to provide its staff greater perspective and insight into customer and policy data.[26] The goal is to help the team drive growth by using this insight to advise the development of new products. Avanade believes this type of insight will lead to a 33 percent revenue growth, based on a survey they commissioned.[27]

Let me share what I think about Avanade's business intelligence approach. I have over thirty years of executive business experience. During my career, I had to make product development decisions with limited information, typically gathered during sales and engineering meetings. Sales people and applications engineers interface with the clients and are a rich source of business intelligence regarding new product development. We also commissioned surveys to determine the demand for a new product. In reality, though, our product development decisions were rarely clear-cut or ensured market acceptance. Return on investment (ROI) data included numerous assumptions, which added little assurance that the decisions were correct. For these reasons, I required that new products have short development cycles, which required the new product be bare bones, and an ROI that ensured we'd recoup product development investment within eighteen months of sales. Even with these restrictions, at least half of all new products fizzled. However, by comparison, without this approach close to 90 percent of new products ended up being duds. Competitors that had long product development cycles and ROIs that allowed product development costs to be recouped in five years often found themselves out millions of dollars and unable to effectively compete. I found it was better to rush products to market and refine them as market demand necessitated. You can argue with the approach, but it worked such that the new products that succeeded in the market covered the development costs of those that failed and still yielded a profit after eighteen months.

Avanade's AI business intelligence algorithm aggregates customer data, taking out the subjective feedback from sales, engineering teams, and any human sources of information. Had it been available, I would have used it.

In general, I don't think current AI business intelligence tools replace human judgment or guarantee success. However, they enable judgments that are more informed. This leads to higher quality timely business decisions and increases the probability that those decisions will be correct.

4. SECURITY

Security is an umbrella term. It can encompass security at the level of an individual, such as reducing credit card fraud, or it can apply in a national context, such as preventing terrorist attacks. Artificial intelligence and machine learning are affecting security at all levels and in all contexts, but the AI applications often generate privacy concerns. For example, to prevent terrorism the government must monitor the activity of a wide spectrum of people, many of whom are not terrorists. Thus, the ethical use of AI in security is a significant question and represents a new frontier for ethics. This is not unusual; technology generally advances faster than our ethical control of the technology. This raises a question: How do we use AI to enhance individual and national security without violating individual rights or the US Constitution? There is no simple answer. As of this writing, it appears AI-based security applications will compromise some portion of our rights to privacy in the interest of national security. How we draw the line is still being debated in the courts. The question becomes even more serious when we discuss autonomous weapons, which on their own can make life-and-death decisions. The severity of this question is enormous and we will discuss it fully in a later chapter. For now, let's consider two examples. The first will relate to individual security and the second to national security.

Identify Theft—According to the American Institute of CPAs, about one in five Americans were the target of identity theft in 2015.[28] Just a decade back, it was common for the credit card company to call you and inquire whether the purchase just charged to the credit card was your purchase. However, today the nature of the call from the credit card company is different. The credit card company calls to notify you that someone attempted to fraudulently charge your card, and they notify you that your account is frozen. How is it possible for the credit card company to know instantly that you did not swipe the card? The answer is machine learning. The credit card companies have algorithms that track your purchasing behavior and numerous other factors, and they are able to predict with uncanny accuracy whether you are making a specific purchase. However, even with these safe guards, identity fraud increased 16 percent in 2016 versus 2015.[29]

Cyberattacks—Almost everything in modern civilization depends on computers and the internet, and cyberwarfare (i.e., attacking of information systems for strategic or military purposes) is a clear and present danger. The US Department of Defense recognizes this threat to national security. In 2009, the United States created the US Cyber Command (USCYBERCOM) at the National Security Agency (NSA) headquarters in Fort George G. Meade, Maryland.[30] According to the US Department of Defense (DoD), USCYBERCOM, plans, coordinates, integrates, synchronizes and conducts activities to: direct the operations and defense of specified Department of Defense information networks and; prepare to, and when directed, conduct full spectrum military cyberspace operations in order to enable actions in all domains, ensure US/Allied freedom of action in cyberspace and deny the same to our adversaries.[31]

In simple terms, USCYBERCOM has a mission to defend the United States against cyberattacks, as well as offensively engage in cyberwarfare. Let's discuss the role AI plays in cyberattacks, using computer viruses as an example.

Sophisticated computer viruses use advanced AI to penetrate a target computer system. The AI is required to disguise the program such that it can penetrate the computer's firewall and continue to evade detection. Malwares and other viruses, for example, change their code dynamically to penetrate firewalls and prevent detection from antivirus software.

On the flip side, antivirus programs utilize machine learning and AI for virus detection. Antivirus scanners detect computer viruses by looking at the code and behavioral patterns to infer if a program is a virus. AI behavior–based scanning can work against highly sophisticated viruses, without requiring a definition of the virus in the antivirus database.

5. FINANCE

The seven leading US commercial banks are strategically investing in AI applications to better serve their customers, improve performance, and increase revenue. The future of finance will depend on emerging fintech (i.e., a contraction of the phrase "financial technology") and AI applications, and will ultimately determine competitiveness among the banking giants.[32]

Large banks process huge volumes of data to generate financial reports and satisfy regulatory requirements. Although these processes are mostly standardized and formulaic, they still require numerous employees to complete. The standardized and formulaic aspect of the processes makes them ideal candidates for robotic process automation (RPA). However, banking challenges arise during the processes that fall outside the scope of RPA. For this reason, banks combine machine learning with RPA. Given these benefits and to avoid disruption, large banks are investing heavily in AI and related technologies, as well as recruiting and developing the talent to work effectively with AI.[33]

Corporations and large organizations are also taking advantage of AI to transform their central functions in finance, including intercompany reconciliations, the quarterly close, and issuing earning reports. AI is also seeping into more strategic corporate functions, including financial analysis, asset allocation, and forecasting. The significant benefits of using AI to assist in corporate finance are accuracy and speed.[34]

Artificial intelligence can help banks and corporations improve operational efficiency, but banking still requires humans to make strategic decisions.

6. THE INTERNET OF THINGS (IOT)

The buzz phrase "the Internet of Things" refers to connecting any device to the internet and to other devices, from thermostats to washing machines, but it excludes PCs, tablets, and smartphones. Gartner, a management-consulting firm, asserts that by 2020 there will be over 26 billion connected devices.[35] The new rule appears to be: If it can be connected to the internet, it should be connected. This raises a question: Why make all these connections? To address that question, let's look at the 2013 definition of the Internet of Things by the Global Standards Initiative:

> A global infrastructure for the information society, enabling advanced services by interconnecting (physical and virtual) things based on existing and evolving interoperable information and communication technologies.[36]

The purpose, based on this definition, is to enable advanced services. Let's take some simple examples to illustrate this. You're shopping at the supermarket, using a shopping list on your phone. Your smart kitchen notices you omitted "salt" and automatically adds that to your shopping list. Here is another simple example. You have a date to meet your wife for dinner, but she is in heavy traffic. Her smart car sends you a text message that she is going to arrive late and provides an estimated arrival time. The IoT is also applicable to larger systems, such as a city's transportation network, which could improve efficiency. For example, if no one is waiting for the train at the next station, the train automatically omits stopping at that station.

The IoT allows for endless opportunities and connections to take place and for those connections to serve us in ways we can't think of or fully understand today. However, while there seems to be unbridled enthusiasm to connect everything to the internet, an important point is not getting sufficient attention—namely, that our reliance on the internet is moving to dependence. We are rapidly implementing the Internet of Things without addressing the profound security challenges involved.[37] In our train example, what happens if someone maliciously hacks the IoT connection between the train and the station while passengers fill the station waiting for the train? The hacker could send an instruction for the train to omit stopping. What if we're not talking about a train, but emergency response vehicles, like EMTs? The hacker could instruct the EMTs to respond to the wrong address. Now, a person's life or death is in the balance. This raises a question: Should the IoT be regulated? As of this writing, the Federal Trade Commission is declining to regulate the IoT. Is this decision wise? You be the judge.

7. WEARABLES

Let's start by understanding "wearables." According to a definition by Techopedia:

> A wearable device is a technology that is worn on the human body. This type of device has become a more common part of the tech world as companies have started to evolve more types of devices that are small enough to wear and that include powerful sensor technologies that can collect and deliver information about their surroundings.[38]

Typical application for wearables includes tracking a user's vital signs, data related to health and fitness, and a user's location. Examples include:

- Apple Watch Series 3—Using the features of the Apple Watch Series 3 you can, based on the specific iPhone model: "Stay connected when you're away from your phone. Make calls and send texts with just your watch. Stream 45 million songs with Apple Music right from your wrist. Ask Siri to set a reminder, send a calendar invitation, or give you directions, all without your phone. Leave your phone at home and still get alerts from your favorite apps. Take your workouts further. Stay in touch while you work out. See how far and high you go. Track a casual walk, an intense ride, and everything in between."[39] As of this writing, prices vary from basic models costing $100–$200 (US) to more sophisticated models costing $300–$400.

- AI Hearables—Let's start by defining AI hearables. According to the Everyday Hearing website, a hearable is "a wireless in-ear computational earpiece. Essentially you have a microcomputer that fits in your ear canal and utilizes wireless technology to supplement and enhance your listening experience."[40] The ultimate goal is to have the complete functionality of a hearing aid and computer in a small device that fits within the ear. However, the technology is far from mature, and we are a long way from the ultimate goal. However, with progress in nanoelectronics and nanosensors, the advent of having internet connectivity citywide in major cities, the ultimate goal appears approachable. The big question is when, not if.

I would be remiss if I did not mention that the wearables market is in its infancy and to date has not lived up to its hype. Much will depend on how other technologies progress, such as nanoelectronics and nanosensors, as well as AI technology. Given what I know about artificial intelligence, nanoelectronics, and nanosensors, I judge we will reach the ultimate goal of AI hearables within a decade or less. In the interim, I expect AI hearables to offer more functionality with each new generation. I also expect the Apple Watch to mature in a similar manner. Eventually, the Law of Accelerating Returns will make the technology compelling and affordable to a wide spectrum of the population.

8. PERSONAL ASSISTANTS/PRODUCTIVITY

It may appear odd that "personal assistants" and "productivity" are in the same category. The reason is simple. The goal of an artificially intelligent personal assistant is to increase the productivity of its user. In fact, the device, commonly categorized as an intelligent personal assistant (IPA), does almost the same tasks as a human personal assistant, including scheduling meetings, booking travel, managing your receipts, taking notes, taking dictation, and completing repetitive sales tasks.[41]

In decades past, only the executive elite had their own human personal assistant. This allowed the executive to focus on only the tasks he could do and left the routine tasks to his personal assistant. When I first entered industry, setting up a meeting was a chore. I would normally have to call each person and find times that were open. Eventually, I would succeed in finding a time that was open to all. Even today, without using an intelligent personal assistant device, the average meeting requires at least three emails before finding a time convenient to all. That is an expensive waste of time, since the people setting up the meeting are knowledge workers (i.e., workers that deal in information, as opposed to physical labor), typically earing higher than minimum wage salaries.

Fortunately, in the late 1990s, IPDs (i.e., Intelligent Personal Devices) began to make their appearance. At the time, the IPDs were separate from your cell phone. All that changed with the introduction of the Apple iPhone on June 29, 2007 and the media frenzy that followed.[42] Prior to that, other smartphones included email functionality and little else. I can remember to this day seeing pictures of Steve Jobs announcing the iPhone at the Macworld convention and thinking: What's so special about the iPhone? As a senior executive of Honeywell, I already had a phone capable of surfing the internet, sending and receiving email, and of course making calls. My smartphone included a miniature keyboard, as did all smartphones prior to the introduction of the iPhone. I also had a human personal assistant and an IPD. With a background as a physicist, I was a techie, but I still found the smartphone's functionality challenging. Although I did not rush out to buy one of the early model iPhones, my sons did. They demonstrated the capability of their new iPhones, which made my smartphone appear feeble. For one thing, the iPhone did not have a miniature keyboard with navigation buttons. Instead, it had a touch screen keyboard and controls, when the application required them. I finally recognized what Steve Jobs accomplished. In Steve Jobs's own words he merged a "widescreen iPod with touch controls," a "revolutionary mobile phone," and a "breakthrough Internet communications device."[43] The iPhone, like a computer, had numerous software applications, which in the early years grew in number exponentially. While the idea of a "smartphone" was not new, no company had a phone offering the functionality of the iPhone. Its invention changed the world. Today, most cell phone manufacturers offer a smartphone that rivals the iPhone, and most people own a smartphone. Eventually, smartphones made the iPod and IPD devices obsolete. The iPhone firmly established Apple as a consumer electronics giant. As I write, I am listening to my iPhone 7 play music on a Bluetooth speaker. I can use voice commands and its Siri application will respond in natural language. Upon request, it will provide the weather forecast, list my upcoming meetings, or schedule a meeting reminder.

Most people in my generation can remember the original *Star Trek* communicator. That device was science fiction. The writers of *Star Trek* could have assigned any capabilities they could think of to it. However, it simply served as a communicator and a location device that could be "locked on" to beam up Captain Kirk and other members of the *Star Trek* crew. However, you never saw Captain Kirk ask it questions or watch a movie on it. The iPhone actually has more functionality that the *Star Trek* communicator.

Like many smartphone owners, I rely on my iPhone to keep me on schedule, provide weather alerts and driving directions, send text messages and email, and take pictures. Some people do even more with their smartphone, via the enormous number of applications available for it.

According to research by Frost & Sullivan, a business consulting company, employees say smartphones boost productivity by 34 percent, gaining fifty-eight minutes of work time and fifty-eight minutes of personal time each day on average.[44]

9. E-COMMERCE

AI is transforming e-commerce. Gartner, a global research and advisory company, predicts AI will handle 85 percent of customer interactions by 2020.[45] However, most of us are experiencing that today. For example, if I call Walgreens to refill a prescription, the entire conversation consists of a natural language interaction with their artificial intelligence application. It has gotten to the point that it is nearly impossible to detect that the inter-action is with a computer application. This is not unique to Walgreens. As mentioned, it is becoming a widespread practice. It is cost effective and efficient to enable an AI computer application to handle the e-commerce function. Let's examine this in more detail.

AI is able to analyze big data (i.e., vast data sets) faster and more effi-ciently than a human. It can rapidly identify clusters and patterns in the information, including similarities between customers, past purchasing behavior, and credit checks. This means it can personalize offers to a spe-cific customer. This is critical to the success of online stores, like Amazon, which rely on AI algorithms to increase conversions (i.e., sales).

AI is also able to act like a virtual sales assistant. It can send you an alert when a product or service you researched online becomes available or goes down in price. For example, if you add a product to your Amazon wish list, Amazon will alert you to availability and price drops. The same holds true for many travel sites that send you alerts regarding air travel availability and price drops. In the near future, by about 2020, I expect the virtual sales assistant to become a personal buying assistant. When instructed via its program, instead of sending an alert, it will purchase the product or service for you at the best price available at a time that fits your schedule.

Chatbots, AI algorithms that interact with customers via natural lan-guage, are becoming more intelligent.[46] Going beyond simple formulated question-and-answer systems, chatbots enrich the customer buying experi-ence. For example, a chatbot connected to a company's delivery process can provide the customer a real-time product delivery status. This person-alized product delivery update improves customer satisfaction. Chatbots are the wave of the future in that they can provide:

- Customer service that is scalable—This means not waiting for the next company representative or being asked to call back at another time.
- Customer intelligence—This means the chatbot is more likely to sat-isfactorily address your enquiry faster than a human.
- Ease of use—This means you can interact with the chatbot using natural language instead of typing text.
- Personalization—This means the chatbot will utilize big data to provide a personal buying experience that causes a customer to feel they are interacting with a "friend," not a company.

The key strength of chatbots lies in their ability to use AI to analyze big data to address customer enquiries in real time and to learn from that interaction. Humans, even using traditional customer relationship management (CRM) databases, inventory databases, etc., are slower, and in the long term more expensive, than AI chatbots. For example, IBM is using its Watson to make correlations between structured and unstructured data, to determine when and what products to order (i.e., inventory management), and even to determine if a product needs to be discounted to be competitive.[47]

In general, using AI in e-commerce makes perfect sense, but it won't always work. Human talent is required in specific areas that AI is unable to address, including:

- Applying knowledge that is not contained in the company's databases but exists in the minds of the company's employees. For example, seasoned sales personnel can handle irate customers, reading verbal clues such as voice pitch, to defuse situations that would otherwise result in a lost sale or a lawsuit.
- Being the voice of the company when a customer insists on talking to a real person.
- Building a personal relationship with a customer, where trust is an important factor in making the sale.

10. ROBOTICS

When we make a robot artificially intelligent, we get a machine that is able to perform functions that normally require a human, and the implications of that capability are enormous. However, let us understand this subject area by looking at each component separately.

Artificial Intelligence (AI) is an umbrella term that implies the use of a computer to replicate intelligent behavior. Consider this: We normally consider the game of chess to require a high level of human intelligence. Yet a typical smartphone with a chess algorithm is able to outplay most humans. However, the smartphone's chess algorithm is only intelligent in this one function. For example, it cannot play checkers, which is a less sophisticated game that many people learn as children. The smartphone can replicate a number of functions that equal or exceed human intelligence, but overall it is less intelligent than a human. It lacks artificial general intelligence. Some AI researchers term artificial general intelligence strong AI. In general, no computer in existence exhibits artificial general intelligence (i.e., none pass the Turing Test). Today, research in AI focuses on the development of algorithms that perform intelligent behavior with minimal human intervention and are even able to learn from experience. Some AI researchers refer to these algorithms as "smart agents." We discussed many of them in the prior sections.

Robotics is a technology that focuses on the design, construction, operation, and application of robots. The term "robot" refers to a machine capable of performing actions automatically. As a branch of technology, it is over a thousand years older than the technology of artificial intelligence. In fact, the earliest account of a robot dates back to the fourth century BCE, when the Greek mathematician Archytas of Tarentum developed a mechanical bird he termed "The Pigeon," which was propelled by steam. Most people are aware that car manufacturers make extensive use of robots, which perform repetitive actions that most human workers would find boring. The US military and police forces use robots for dangerous tasks, such as defusing bombs.

Early robots were a form of automation, representing a machine that by construction could do a task normally requiring a human. However, the scope of early robots was limited. By construction, they could perform a single task. To do another task required modifying the robot or building a new robot for that task. With the advent of computers, that changed.

In 1954, American inventor George Devol developed the first digitally operated programmable robot, which he called "Unimate." In 1956, Devol and American physicist, engineer, and entrepreneur Joseph Engelberger founded Unimation, Inc., the first robotics company. In 1960, Unimation sold the first Unimate to General Motors (GM). General Motors installed the Unimate in 1961 in a plant in Trenton, New Jersey. Its programming enabled it to lift hot pieces of metal from a die-casting machine and stack them in preparation for the next assembly operation. Devol's invention of the Unimate and his patent for the first digitally operated programmable robotic arm is seen by many as the foundation of the modern robotics industry.[48] Programmable robots went beyond simple automation. It became possible to program one robot to perform multiple tasks, in much the same way we cross-train humans to do multiple tasks. In the Unimate, we see the first glimpse of rudimentary AI.

Although many others developed programmable robotic arms, they continued to represent rudimentary AI. That changed in 1970 when the Stanford Research Institute (now SRI International) developed the first mobile robot capable of reasoning about its surroundings, which they called Shakey.[49] Shakey's design included multiple sensors, including TV cameras, laser rangefinders, and "bump sensors," which enabled it to navigate on its own.

If we fast-forward to the present, artificially intelligent robots are becoming commonplace in every facet of modern civilization, from the living room vacuum cleaner robot, Roomba, released in 2002 by the company iRobot, to the US military's MQ-1 Predator drone, made by General Atomics.

11. EDUCATION

Although it is possible to trace the application of AI in education back to the early 1980s, it is only within the last decade that AI has been profoundly changing education.

In generations past, education relied on textbooks and human instructors. Typically, students attended a class, characterized by a room filled with desks, at a specific time to learn a specific subject. The instructor typically taught the material in the textbook, at times going beyond what the textbook included and at other times covering less than the textbook included. During class, a student typically had the opportunity to ask questions, but at home the student had to rely on the textbook for answers. This was especially true prior to the introduction of the internet. If the instructor assigned the student a subject to research, the student typically needed to make their way to a library, where the librarian would help them find the material they sought. In the past two decades, the internet has provided another avenue for research and addressing questions. Personal computers have also made changes to the education process. Using numerous applications, a student is able to develop reports that rival those found in scholarly articles, complete with footnotes, endnotes, and references. As profound as these changes are, the real impact of AI on education is just beginning to emerge.

AI's digital, dynamic nature encourages student engagement beyond textbooks or the classroom environment. Over the last decade, artificial intelligence applications, known as Intelligent Tutor Systems (ITS), provide immediate customized instruction and feedback to students.[50] ITS are also able to track the student's mental steps during problem-solving tasks and use the information to determine the student's level of understanding and resolve any misconceptions. ITS can also assign learning exercises at an appropriate difficulty level for a specific student. These systems provide the benefits of one-to-one tutoring, and some approach the effectiveness of expert tutors. Examples of these intelligent tutor systems include Carnegie Learning, Tabtor, and Front Row. A comparison of learner outcomes using Intelligent Tutor Systems to using other instructional methods indicates that learning from ITS leads to higher student-learning scores.[51]

In addition to Intelligent Tutor Systems, which are able to provide one-on-one tutoring, another application, termed enhanced crowd-sourced tutoring, is enabling social networking that helps millions of students collaborate. This is similar to the age-old practice of getting help from classmates. A notable example of enhanced crowd-sourced tutoring is Brainly, which uses over a thousand moderators as gatekeepers to scrutinize questions and validate answers users put on the platform.[52] Brainly also uses machine-learning algorithms to automatically filter spam and low-quality content, which enables moderators to focus on providing quality services to students.

Another application of artificial intelligence in education is deep learning systems, which read, write, and emulate human behavior to deliver custom content. For example, Content Technologies enables educators to assemble custom textbooks from their course syllabus.[53]

The examples above make it clear that AI can enable a faster and deeper learning than traditional methods. AI algorithms will be capable of replacing human instructors completely in one to two decades. I make this assertion because AI is on a course to equal human-level intelligence in roughly the same timeframe. Schools and classrooms, as we now know them, will also change considerably as AI takes the lead in educating students. However, please notice I used the word "capable" not "will" in describing AI replacing human instructors. I carefully chose this wording because the investment in AI algorithms, databases, computers, and other associated hardware and software will mitigate how quickly AI replaces humans as instructors. Given even this small glimpse into AI's impact on education today, though, especially as related to its effectiveness in providing deeper and faster learning experiences, society will need to move with the technology. Earlier, we discussed the worldwide shortage of over 7 million physicians, nurses, and other health workers and the enormous time and expense it currently takes to train them. AI can change this. Conceivably, once a learning institution makes the investment in the appropriate hardware and software, it will reduce the time and cost required to train medical professionals. Going beyond this, I can envision a future in which each student has their own computer with algorithms designed to teach specific courses. This would make learning on demand a reality. I can also envision a future where each person has an artificially intelligent personal assistant that assists the user in lifelong learning. Imagine a doctor who encounters a patient with a rash that the doctor has previously never encountered. Imagine the doctor using his artificially intelligent personal assistant, incorporated in a smartphone, to take a photograph of the rash. The artificially intelligent personal assistant queries big data, via the internet, identifies the rash, and recommends treatment options. This may border on science fiction today, but it is likely to be science fact in the coming decades.

THE MAJOR POINTS

This chapter provides only an overview of the numerous commercial, industrial, and medical AI applications. Only representative algorithms and examples are included in each of the eleven categories. It would be possible to devote an entire book to these applications, but our focus here was to make two points:

1. **AI is an important element in almost every aspects of modern society and our reliance on AI is moving to**

dependence. On our current course, by the third quarter of the twenty-first century, modern society will likely be unable to function without AI. While AI is greatly enriching our lives today, our growing dependence on it is concerning. Any disruption to AI technology could threaten the survival of our species. I offer this as food for thought for now, but this is an important point that we will revisit in a later chapter.

2. **Modern society is unaware of its growing dependence on AI.** As we discussed, this blindness to the increasing role of AI is the AI effect. Unfortunately, this lack of awareness in no way hinders AI progress, but it does result in a laissez-faire attitude toward it. For example, there is no legislation regulating the development or application of AI, including its use in warfare. This could also threaten the survival of our species. Some countries, like Russia, are already fielding autonomous weapon systems.[54] This means they are allowing artificially intelligent machines to strike without human control and make life-and-death decisions with regard to humans. This has enormous implications regarding the future of warfare. Will autonomous weapons ignite a world war? What issues will humanity face at the point of the singularity, when AI greatly exceeds the cognitive performance of humans in virtually all domains of interest? Will superintelligences view humanity as a threat to their existence, given our war-faring history and our malicious behavior of unleashing computer viruses? Will they use our own weapons to eradicate us as we would an infestation of killer bees?

The AI effect combined with the numerous beneficial AI applications results in a humanity that is blind to the malevolent side of AI. However, AI has a dark malevolent side, even if we don't see it. If this sounds scary, it is because the reality is scarier than the stuff usually found in science fiction.

*A nation's ability to fight a modern war is as good as its techno-
logical ability.*

—Frank Whittle, inventor of the turbojet engine

CHAPTER 3

I, ROBOT AM DEADLY

There is a new arms race. The United States, China, and Russia are placing artificial intelligence at the center of their new weapons strategy. Every country is being secretive regarding its development and deployment of artificially intelligent weapons. However, it is possible to gain insight into even the most secret areas by using a time-proven technique—namely, apply the old adage "follow the money."

The countries with the top three largest defense budgets are, in descending order, the United States, China, and Russia. Specifically, the United States has the largest military budget in the world. In 2016, the US Department of Defense spent slightly over $611 billion (US) on defense, 3.3 percent of the US gross domestic product (GDP). In comparison, China, which has the second largest budget, is spending slightly more than $215 billion (1.9 percent of their GDP), roughly a third of what the United States spends. Russia is a distant third at slightly over $69 billion (5.3 percent of their GDP), roughly a little over 11 percent of what the United States spends.[1]

Given these defense budgets, a person may rush to conclude that the United States would reign supreme in all aspects of warfare. Unfortunately, that is not how it works. Both China and Russia understand that they cannot match the United States one-to-one in every aspect of warfare.

In recent years, increased spending by China and Russia on modernization is closing the military leadership gap the United States has enjoyed. If we look closely at China's military modernization over the last fifteen years, we can see that it includes ballistic missiles, air defense, aircraft, electronic warfare, and naval vessels. China is not trying to match US military might across the board. Their objective is greater control over the Asia-Pacific region, especially control over their near seas. Their focused objective does not require military parity with the United States, whose global mission is to ensure freedom of navigation and commerce, including the hotly con-tested Asia-Pacific seas. With this limited goal, China only needs to be a viable threat to the United States in the Asia-Pacific region. Given their modernization over the last fifteen years, especially with regard to their naval and ballistic capabilities, China appears to have anti-ship missiles capable of destroying US aircraft carriers and missiles capable of attacking air bases in the region or even top-tier US fighter aircraft, like the F35. Their cyber systems appear able to disrupt US logistics and communi-cations. In light of this, it is reasonable to conclude they are achieving their objective. Russia, like China, is also seeking to develop new military technologies to undermine US capabilities in Europe and Asia. In past conflicts, such as Desert Storm, the United States did not face adversaries capable of destroying its aircraft carriers and air bases or challenging its ability to dominate the adversary's air space. This asymmetrical aspect of warfare creates a new problem for US military planners, and challenges the United States' ability to project force.[2]

An important focus in China and Russia's military investments is to gain asymmetrical advantages in the application of artificial intelligence to weapons, commonly termed autonomous weapons. This raises a question: What is driving this trend?

China and Russia have concluded that artificially intelligent weapons are critical to gaining an asymmetrical edge over the United States. Both nations, as well as the United States, understand that technology is a key factor in the next generation of warfare. However, let's understand the spe-cific driving force for each.

Chinese researchers are fully engaged in AI research, which is enabling numerous commercial advances in artificial intelligence among Chinese companies. For example, Andrew Ng, the leader of a Silicon Valley laboratory for the Chinese web services company Baidu, led a team that developed an AI algorithm in 2015 that surpassed human-level Chinese language recognition.[3] A year later, Microsoft researchers proclaimed that their company had created software capable of matching human skills in understanding speech. While Microsoft garnered the media spotlight, Baidu quietly stood in the shadow, knowing they were a year ahead in that AI technology. This is not just an anomaly. Another example that went largely unreported in the United States is Iflytek, an artificial intelligence company that focuses on speech recognition and understanding natural language.[4] On the world stage, it won international competitions in speech synthesis and in translation between Chinese- and English-language texts. China's military strategists know that much of the technology that once came only from the US government and its suppliers now comes from the commercial sector. The same companies that make high-end gaming software and computers are the leaders in AI technology, not the US government labs. This means that if the US military intends to continue superiority in autonomous weapon technologies that require artificial intelligence, the United States will need to restructure its autonomous weapons procurements to include procurements from commercial corporations with AI expertise. To that end, the Pentagon established the Defense Innovation Unit Experimental facility (DIUx) in Silicon Valley to streamline US government contracting practices to accommodate the faster more fluid style of Silicon Valley[5]. China also understands this paradigm shift. Baidu drove home this point in 2017 when Qi Lu, a Microsoft artificial intelligence specialist, left the company to become chief operating officer at Baidu. Lu will oversee Baidu's plan to become a global leader in AI. In summary, China knows that their strong commercial base in AI is a prerequisite to a strong military edge in autonomous weapons.[6]

Russia has limited options. Although Russia may have nuclear parity with the United States, it lags in most other elements of warfare. In particular, Russia has a small population compared to the United States and China. The population of each country (rounded to the nearest whole digit) is:

- China: 1.4 billion
- United States: 319 million
- Russia: 142 million[7]

As far as world standing, China is the most populated country on Earth. The United States is the fourth most populated. Russia is the ninth most populated. This fact weighs heavily into Russia's thinking about automation and artificial intelligence. Their population, among other things, puts them at a disadvantage when it comes to being a major player. Therefore, they have announced a strategy to have a robotic army and deploy autonomous weapons. In addition, the export of weapons plays a critical role in Russia's economy. It accounts for a significant proportion of manufactured and technology-intensive exports. In fact, Russia is the world's second-largest arms exporter after the United States. Russia's armaments industry enables Russia to integrate into the global economy, and it helps preserve Russia's full spectrum of capabilities. Russian leadership knows that their next generation of weapon exports needs to incorporate artificial intelligence to be competitive in the world market.[8]

The United States is aware of the changing landscape in weapons technology and adversaries. In response, on November 2014 the Department of Defense announced the "Third Offset Strategy."[9] Let's discuss what this means.

An Offset Strategy uses technology to overcome military advantages of top adversaries. Its principles are twofold. First, it seeks to have the military technological might to win a war if necessary. Second, and most importantly, it seeks to have enough technological military capability to deter one. To understand Offset Strategies, let's examine each of the US Offset Strategies in the timeframes they were implemented.

1. The First Offset Strategy: In the 1950s, President Eisenhower emphasized technological superiority in nuclear weapons as a defense and deterrence to aggression by the Warsaw Pact. This enabled the United States to avoid the larger expenditures necessary to conventionally deter the Warsaw Pact from initiating hostilities. The First Offset Strategy continued to be effective through the 1960s.

2. The Second Offset Strategy: From 1975 to 1989, the United States emphasized technological superiority to offset numerical advantages held by US adversaries and restore deterrence stability in Europe.[10] In Europe, Warsaw Pact forces outnumbered NATO forces three to one. In light of this, Secretary of Defense Harold Brown, under

President Jimmy Carter, emphasized new intelligence, surveillance, and reconnaissance (ISR) platforms, improvements in precision-guided weapons, stealth technology, and space-based military communications and navigation.[11] In concrete terms, this resulted in the Airborne Warning and Control System (AWACS), the F-117 stealth fighter and its successors, modern precision-guided munitions, and the Global Positioning System (GPS). This led military strategists and the public during this timeframe to refer to the US capabilities as "smart weapons," which heavily relied on leadership in integrated circuit and sensor technology.

3. The Third Offset Strategy: The Third Offset Strategy was established in 2014 to enable the United States to maintain a military advantage in the face of the rapidly advancing military technological capabilities of China and Russia. Let me put this in perspective. In the late 1980s, the United States began to find it harder to maintain technological leadership based on advanced integrated circuits and sensors. Inexpensive and universally available integrated circuits and sensors, along with the hardware for their manufacture, made it difficult for the Pentagon to control the technological progress of adversaries like China and Russia. Countries throughout the world began to manufacture advanced integrated circuits and sensors. That technology was no longer the sole provenance of military and advanced corporate laboratories. New technologies increasingly came from consumer electronics firms. This raised a serious question: What Offset Strategy could the United States employ against adversaries with access to advanced integrated circuits and sensor technology? In November 2014, Secretary of Defense Chuck Hagel, under President Obama, announced the Third Offset Strategy. It called for establishing a long-range research and development planning program to target several promising technology areas, including artificial intelligence, robotics, and miniaturization. This meant the US Department of Defense (DoD) would develop and deploy weapons exploiting robotics and system autonomy, miniaturization (including nanoweapons), big data, and advanced manufacturing. In addition, there was a shift in the DoD development thrusts to improve the US military's col-

laboration with innovative private sector enterprises.[12] In October 2014, the Center for Strategic and Budgetary Assessments released a report delineating the components of the Third Offset Strategy, prior to Secretary Hagel's announcement. The report emphasized the development of next-generation power projection platforms, including unmanned autonomous strike aircraft, the acceleration of the LRS-B (long-range strike bomber), unmanned underwater vehicles, and strategies to reduce the United States' vulnerability to the loss of space-based communications. I'd like to emphasize that no announcements include the term "nanoweapons." Instead, the announcements use the term "miniaturization." However, there is no doubt that nanoweapons, a military capability in which the United States enjoys a substantial lead, is a critical component to the Third Offset Strategy. Nanoweapons, along with other elements of the Third Offset Strategy, will be explained in depth shortly.

According to Deputy Defense Secretary Bob Work, the Third Offset Strategy will be conventional and characterized by "pacing competitors" like China and Russia.[13] Work asserted,

> Our pacing competitors have put a lot of money in counter-network operations because they know how powerful our battle networks are, so they spend a lot of money on cyber capabilities, on electronic warfare capabilities and on counter-space capabilities because our space constellation is a very important part of our ability to put these battle networks together.[14]

Work explained that the Strategic Capabilities Office, now a part of the Third Offset Strategy, will utilize systems that the DoD already has invested in significantly, and transform or repurpose them in ways the world has never seen or encountered. He clearly asserts that artificial intelligence and autonomy is central to the Third Offset Strategy, as is a focus on having an advantage at the operational level of war, "because historically . . . having that advantage is the surest way to underwrite conventional deterrence."[15]

In summary, the United States now plans to retain their worldwide military advantage via artificial intelligence and robotic weapons, as delineated in the Third Offset Strategy, and they intend to do it, in part, with technology that is widely available on the open market. This raises a question: What are the challenges to successfully implementing the Third Offset Strategy?

The challenges to successfully implementing the Third Offset Strategy are significant. Would it surprise you to learn that in 2016 the Islamic State (ISIS) in Iraq launched an off-the-shelf commercial drone booby-trapped with explosives to kill two Kurdish soldiers and wound two French paratroopers? Unfortunately, it is true. In 2016, hobby-level drones found application in combat operations from Iraq to Syria to the Ukraine.[16]

ISIS's drone attack is only one example of a global trend. Commercially available new technology spreads globally more quickly than in previous decades, which narrows the US military's technology gap. America's technology advances can at times be insufficient to compensate for adversaries' progress. For example, in cyberspace, Russia and China can now match America's capabilities. Although adversaries may not have parity with the United States across the board, they can pose a significant threat in their area of influence, such as China in the Asia-Pacific. We are coming dangerously close to a point where China's anti-access area denial weapons may make operating in the South China Sea risky for the United States.[17]

It is becoming obvious that the United States' technological advancements will last only a few years before adversaries develop countermeasures or similar capabilities. Therefore, the Pentagon should proceed with weapons development similar to the way traditional high-tech businesses, such as Apple or Google, constantly innovate to compete on the world stage. In addition, the United States must prepare for a scenario where an adversary is able to take out our high-tech capabilities, such as our GPS, surveillance, and communication satellites. In simple terms, we must learn to fight and win against our most capable adversaries using only our low-tech capabilities.

The United States' Third Offset Strategy emphasizes technological superiority in artificial intelligence, robotics, and miniaturization (including nanoweapons), and we can gain a deeper understanding of the implications of this strategy by examining each element independently.

Artificial Intelligence in Weapons

The concept of adding artificial intelligence to weapons suggests "autonomous weapons." The US military categorizes autonomous weapons into defensive autonomous weapons and offensive autonomous weapons. Autonomous weapons may find use in nonlethal missions, such as surveillance. Lethal autonomous weapons (LAWS) refer to weapons designed to select and attack military targets without human intervention. The machine (LAWS) makes life-and-death decisions. In addition, autonomous weapons may operate on land, in the air, on water, underwater, and in space. It is important to differentiate autonomous weapons from remotely controlled military robotic systems, such as US Air Force drones. Although US Air Force drones do have some autonomous features, like autopilot, they are not autonomous weapons. The release of a weapon from a US drone requires a human decision. Let us be clear, an autonomous weapon is able to act on its own. For example, a lethal autonomous weapon would choose its own targets, including human targets, in support of its overall mission. This is highly concerning and the United Nations is addressing banning lethal autonomous weapons, colloquially termed "killer robots." As of September 29, 2017, no United Nations ban of LAWS exists. The United States, China, and Russia currently deploy autonomous weapon systems with various degrees of human control. For example, the US Navy Phalanx Close-In Weapon System uses a radar-guided gun that can autonomously identify and attack anti-ship missiles. The reason for eliminating human control is the need for rapid response. Russia and China field similar systems. In general, the United States' current policy is "Autonomous . . . weapon systems shall be designed to allow commanders and operators to exercise appropriate levels of human judgment over the use of force."[18] It may appear the words "autonomous" and "appropriate levels of human judgment over the use of force" don't belong in the same sentence. While it is true that the United States only wants to field semiautonomous weapons, it recognizes that some conflict situations will occur faster than humans can respond. Hence, it deploys the Phalanx. Another such area is cyberwarfare, where attacks can come instantly and completely without warning. In the area of cyberwarfare, the United States allows some elements of its cyberdefenses to be autonomous and considers this an appropriate level of human control. We will discuss this more fully shortly.

Robotics in Weapons

Military robots, incorporating artificial intelligence, can be autonomous, semiautonomous, or remote-controlled robots designed for military applications. In general, the US military is significantly investing in research and development with the goal of deploying increasingly automated systems. For example, the US military is developing autonomous fighter jets and bombers to destroy enemy targets.[19] This is highly attractive because autonomous fighter jets and bombers do not require training a human pilot. Since no humans are involved, autonomous planes are capable of performing maneuvers that are not possible with human pilots (due to high g-forces). In addition, autonomous planes do not require a life-support system and other systems that keep the pilot safe during combat, such as a pilot-ejection parachute. Artificial intelligence is addressing the largest drawback to robotics, namely their inability to accommodate nonstandard conditions. Because of AI's rapid advances in recent years, specifically in the field of machine learning, AI systems are able to recognize complex or subtle patterns in large quantities of data, which enables the system to perform operations as well as or better than humans.

Miniaturization (i.e., nanoweapons)

I am interpreting "miniaturization" to include nanoweapons, which has numerous implications. It may suggest the weapon itself is small, such as a fly-sized drone, capable of entering an adversary's command center to perform surveillance or other nefarious acts, like assassination. It may also suggest the weapon is using nanotechnology, such as nanoelectronics, which commercial companies like Intel and others use to manufacture microprocessors and other integrated circuits. It could also apply to the use of nanometals (i.e. metals controlled or altered on a nanometer scale), such as nanoaluminum, which the United States currently uses to increase the destructive power of their conventional explosives. In my book, *Nanoweapons: A Growing Threat To Humanity*, I developed a methodology to categories nanoweapons. The US military is especially secretive about nanoweapons. They did not categorize them. In order to make it easier for the lay reader to understand nanoweapons, I developed the five following military nanoweapon categories:[20]

1. Passive Nanoweapons—This category relates to any use of nan-otechnology in warfare that has a nonoffensive/nondefensive application but may increase the effectiveness of conventional or strategic (capable of mass destruction) weapons. In many cases, passive nanoweapons will have a commercial, industrial, or medical counterpart.
2. Offensive Tactical Nanoweapons—This category relates to offensive weapons whose nanotechnology components enhance their tactical capabilities.
3. Defensive Tactical Nanoweapons—This category relates to defensive weapons whose nanotechnology components enhance their tactical capabilities.
4. Offensive Strategic Nanoweapons—This category relates to offensive strategic weapons whose nanotechnology components enhance their strategic capabilities. It also includes offensive autonomous smart nanobots.
5. Defensive Strategic Nanoweapons—This category relates to defensive strategic weapons whose nanotechnology components enhance their strategic capabilities. It also includes defensive autonomous smart nanobots.

As far as the Third Offset Strategy, only categories 2 through 5 play a role. We will discuss them as we discuss semiautonomous weapons later in this chapter.

In addition, we need to understand the DoD Directive 3000.09, which provides guidelines for the development and use of autonomous weapon systems.[21]

The DoD Directive 3000.09 represents the first policy announcement by any country on autonomous weapons (i.e., weapons designed to select and engage targets without human intervention). An important phrase to understand is "human intervention." In practical terms, human involvement regarding the use of weapons can take three forms:

1. In-the-loop: human-controlled weapons
   a. Weapons that can select targets and deliver force only with a human command.

b. Robotic weapons that are remotely controlled by a human operator, such as US Air Force drones. The robotic weapons may have some autonomy, such as navigation, systems control, target detection, and weapons guidance, but they cannot attack without the real-time command of their human operator.

2. On-the-loop: human-supervised weapons

a. Weapons that can select targets and deliver force with the oversight of a human operator who can override the weapon's actions.

b. Weapons that can carry out a targeting process independently from human command but remain under the real-time supervision of a human operator who can override any decision to attack.

3. Out-of-the-loop: autonomous weapons

a. Weapons that are capable of selecting targets and delivering force without any human control or oversight.

b. Weapons that can search, identify, select, and attack targets without real-time control by a human operator. We need to qualify this statement further. A landmine is not an autonomous weapon, it is "automated" in that it can automatically detect and attack targets in a restricted, predefined, and controlled environment. A weapon is said to be autonomous when it is capable performing these tasks in an open and unpredictable environment.

While DoD Directive 3000.09 clearly prohibits the US military from deploying autonomous weapons, it includes important exceptions:

> cyberspace systems for cyberspace operations; unarmed, unmanned platforms; unguided munitions; munitions manually guided by the operator (e.g., laser- or wire-guided munitions); mines; or unexploded explosive ordnance.[22]

These exceptions will be significant when we examine the role of the US Army and other service branches in cyberwarfare.

With the above foundation, we are now ready to discuss the application of AI, robotics, and nanoweapons technology to semiautonomous weapon systems. We will start by discussing those developed or in development by the United States, categorized by military branch. Each category will seek only to present semiautonomous weapons (i.e., autonomous weapons deployed with a human in- or on-the-loop) representative of the category. The examples will delineate those deployed as well as those in development.

US NAVY SEMIAUTONOMOUS WEAPONS

You may wonder: Why start with the US Navy? Based on publicly available information, the US Navy has arguably the most sophisticated semiautonomous weapons. For our purposes, I chose not to highlight the larger platforms, such as aircraft carriers and nuclear submarines. It is possible to argue they are the most devastating naval weapon systems, but they are actually the sum of many different weapon systems. For the sake of clarity, we will focus on specific semiautonomous naval weapons and capabilities.

For our first example, let's consider the Aegis Weapon System. In 1973, the US Navy installed the first Aegis Weapon System (AWS), termed the Engineering Development Model (EDM-1), in the test ship USS *Norton Sound*. I recall working for Semiconductor Electronic Memories Inc., now defunct, in 1971, which had a contract to deliver integrated circuit memories for the Aegis system. The manufacture of the integrated circuit memories for Aegis was identical to those we sold on the open market, but represented advanced technology. This illustrates an important point. Military systems typically use the most advanced technology available. They do this because it takes a long time to develop a weapon system, and when deployed the technology is often no longer advanced. It is current.

According to the US Navy, "The Aegis Weapon System (AWS) is a centralized, automated, command-and-control (C2) and weapons control system that was designed as a total weapon system, from detection to kill."[23] AWS is at the heart of the US Navy's ability to fight numerous adversaries at once, on land, in the air, on the water, or under the water. It uses powerful computers, AI algorithms, and radar technology to direct ballistic missiles against incoming targets. Aegis can coordinate the defense of an entire naval surface group. The Navy continues to upgrade Aegis as new technologies become available. Through the 1990s, for example, the Navy developed and deployed the Cooperative Engagement Capability (CEC), which enables combat systems to share sensor data that allows the battle-group units to operate as one and fire on targets at extended ranges using targeting data from multiple platforms.[24] As computer, AI, and radar technology advances, improvements continue to the CEC.[25] According to the website of Lockheed Martin, the company that produces the Aegis:

> Aegis is the most-advanced, most-deployed combat system in the world. The flexible nature of Aegis allows the system to meet a variety of mission requirements. The Aegis Combat System has evolved into a worldwide network, encompassing more than 100 ships among eight classes in six countries—Australia, Japan, Norway, Republic of Korea, Spain and the United States.[26]

Aegis is an excellent example of a defensive semiautonomous weapon system, integrating computer technology, AI algorithms, and radar technology. It does what humans alone cannot do: It aggregates in real time the network of sensor data collected during battle and provides missile defense options to neutralize the threats. As competitor capabilities evolve, the US Navy continues to evolve the Aegis to maintain its ability to neutralize threats. Its wide adoption by worldwide navies speaks to its flexibility and capability. One last point—the Aegis semiautonomous weapon system looks more like a desktop computer than it does a weapon, as figure 1 illustrates.

Figure 1. Pacific Ocean (July 29, 2010): Ensign Angelique M. Clark stands watch as the guided-missile cruiser USS *Cape St. George* (CG 71) tests its Aegis Weapon System. (US Navy photo by Mass Communication Specialist 2nd Class Arif Patani/Released)

Some autonomous weapon systems will more closely resemble computer technology than the image we likely conjure in our mind's eye when we think of weapons.

For our second example, let's consider the X-47B UCAS (i.e., Unmanned Combat Air System) drone. The X-47B UCAS is a demonstration unmanned combat air system built by Northrop Grumman and designed for aircraft carrier–based operations in the US Navy.[27] It is a tailless, jet-powered, blended-wing-body, semiautonomous aircraft capable of aerially refueling.[28] Its intended purpose is to penetrate deep inside an adversary's well-defended territory for purposes of intelligence, surveillance, and reconnaissance (ISR), and for strike missions. Current projections are that the X-47B will enter service in 2023.[29] Originally, the US Navy intended the X-47B to be autonomous, but the Navy is bypassing this politically charged issue by downgrading X-47B UCAS to a semiautonomous system. Unfortunately, this also sidetracks the higher mission goals the Navy envisioned for X-47B, since communications with the X-47B in its semiautonomous role compromises its stealth capabilities. What makes the X-47B semiautonomous? It can take off and land on an aircraft carrier with minimal to no human intervention.[30] It is also aerially refuelable with minimal to no human intervention.[31] However, the release of weapons is still under human control. Even in semiautonomous mode, the X-47B enables the Navy to have a long-range aerially refuelable drone capable of fighting side by side with manned Navy aircraft. Unlike the Aegis Combat System, this robotic done more closely matches our mental image of a semiautonomous weapon, as figure 2 demonstrates.

These are just two examples of the US Navy's push toward semiautonomous weapons. There are many more examples, but these examples accomplish two goals. The Aegis Combat System demonstrates the enormous potential of powerful computers combined with sophisticated AI algorithms. It is the most remarkable naval defense system in the world. I also chose it to demonstrate that semiautonomous weapon systems do not necessarily conform to what our minds envision as a weapon. The second example, the X-47B UCAS, is exactly what most people envision when they think about a robotic semiautonomous drone. The US Navy is only deploying the X-47B in its semiautonomous mode, intentionally suppressing its autonomous capabilities to bypass the legal, political, and ethical ramifications associated with autonomous weapons, which we'll discuss in chapter 7.

Figure 2. Atlantic Ocean (July 10, 2013): An X-47B Unmanned Combat Air System (UCAS) demonstrator completes an arrested landing on the flight deck of the aircraft carrier USS *George H. W. Bush* (CVN 77). The landing marks the first time any unmanned aircraft has completed an arrested landing at sea. *George H. W. Bush* is conducting training operations in the Atlantic Ocean. (US Navy photo by Mass Communication Specialist Seaman Lorelei R. Vander Griend/Released)

US ARMY SEMIAUTONOMOUS WEAPONS

One of the most important roles the US Army is playing in the Third Offset Strategy is its role in cyberwarfare. As we previously discussed, the DoD Directive 3000.09 prohibits the US military from deploying autonomous weapons, but excludes cyberwarfare.

As previously mentioned, on June 23, 2009, the secretary of defense directed the commander of US Strategic Command to establish US Cyber Command.[32] Following this directive, the commander of US Strategic

Command established US Cyber Command (USCYBERCOM) under the National Security Agency (NSA), headquarters in Fort George G. Meade, Maryland, with the following mission:

> USCYBERCOM plans, coordinates, integrates, synchronizes and con-ducts activities to: direct the operations and defense of specified Depart-ment of Defense information networks and; prepare to, and when directed, conduct full spectrum military cyberspace operations in order to enable actions in all domains, ensure US/Allied freedom of action in cyberspace and deny the same to our adversaries.[33]

General Keith B. Alexander, the director of the NSA and chief of the Central Security Service, officially assumed command of US Cyber Command on May 21, 2010.[34] Under General Alexander's command, US Cyber Command reached full operational capability in October 2010.[35] Its personnel are taken from each branch of the service.

Cyber Command uses NSA networks, and the director of the National Security Agency heads Cyber Command. It originally had a defensive mission, but that has changed. It now has both a defensive and offensive mission, which then commander of Cyber Command, General Alexander, made clear in his May 2010 report to the United States House Committee on Armed Services subcommittee:

> My own view is that the only way to counteract both criminal and espi-onage activity online is to be proactive. If the US is taking a formal approach to this, then that has to be a good thing. The Chinese are viewed as the source of a great many attacks on western infrastructure and just recently, the US electrical grid. If that is determined to be an organized attack, I would want to go and take down the source of those attacks.[36]

On December 23, 2016, President Obama signed the National Defense Authorization Act (NDAA) for fiscal year 2017 into law, which elevated USCYBERCOM to a unified combatant command.[37] This recognized the dual-hatted arrangement of the commander of USCYBERCOM and specified it continue in that role until the secretary of defense and chairman of the Joint Chiefs of Staff jointly certify that ending this arrangement will not pose risks to the national security interests of the United States. In July 2017, the Trump administration initiated plans to revamp the nation's military command for defensive and offensive cyber operations, with the goal of intensifying the United States' ability to wage cyberwarfare against the Islamic State group and other foes.[38] These plans call to split Cyber Command from the National Security Agency, with the goal of giving Cyber Command more autonomy and removing constraints that stem from working alongside the NSA.[39] In April 2018, the Senate unanimously confirmed Lt. Gen Paul Nakasone as the commander of US Cyber Command and director of the National Security Agency.[40] Shortly following his confirmation, Lt. Gen Nakasone began preparing a recommendation regarding whether Cyber Command was ready to split from its dual-hat relationship with the National Security Agency.[41] As of July 11, 2018, no recommendation or decision had been publically announced. However, even if Cyber Command separates from NSA, there is still likely to be an enormous amount of interagency cooperation, since the NSA employs three hundred of the country's leading mathematicians and has a supercomputer. Duplicating that capability would be nearly impossible and, to a certain, extent an unnecessary expenditure. As of July 2017, US Cyber Command had over seven hundred military and civilian employees. Each of the military services also had cyber units, and were working to create 133 operational teams composed of approximately 6,200 personnel.[42]

At this point, you may wonder: How real is cyberwarfare? It's as real as any other aspect of warfare, and just as deadly. US leaders express concern over attacks on US electrical grids and DoD information networks. This means the new battlefield includes your living room, as well as areas of armed conflict in places like Iraq and Syria. For example, if hackers were to knock out one hundred strategically chosen electricity generators in the Northeast, the damaged power grid would quickly overload. This would cause secondary outages across multiple states, leaving some without power for weeks. According to an article in the *Hill*:

> A prolonged outage across 15 states and Washington, DC, according to the University of Cambridge and insurer Lloyd's of London, would leave 93 million people in darkness, cost the economy hundreds of millions of dollars and cause a surge in fatalities at hospitals.[43]

Experts agree such an attack would constitute an act of war.[44] This means the cyberattack could escalate into armed conflict.

The threat is real, and the potential damage is enormous. This is why there is serious consideration regarding making Cyber Command the sixth branch of the US military, on equal footing with the Air Force, Army, Coast Guard, Marine Corps, and Navy.[45]

The mission of the US Army Cyber Command, the Army's separate cyber division, is:

> Army Cyber Command directs and conducts integrated electronic warfare, information and cyberspace operations as authorized, or directed, to ensure freedom of action in and through cyberspace and the information environment, and to deny the same to our adversaries.[46]

The Army had thirty cyber teams at full operational capability as of February 9, 2017, and planned to have forty-one by year's end.[47] In short, the US Army, as well as other service branches, are pushing hard to get up to full force. In 2014, Secretary of Defense Chuck Hagel ordered the US military to add six thousand cyber specialists by 2016.[48] As of 2017, the Army Cyber Command made up about a third of the Cyber National Mission Force.[49] The good news is that it is paying off. According to the *New York Times*, military cyber teams are altering ISIS fighters' electronic messages, "with the aim of redirecting militants to areas more vulnerable to attack by American drones or local ground forces."[50]

As is clear from figure 3, the weapons of cyberwarfare are computers and algorithms, as well as personnel highly trained in cyberwarfare.

Cyberwarfare is also taking to the battlefield, literally. The Army is planning to add two cyberdefense specialists to each combat brigade. Their mission will be to protect their combat units' wireless networks on the battlefield and to attack their enemies' wireless networks.[51] Rather than having to go radio silent, which puts our military at a disadvantage, they plan to jam the enemies' communication network with electronic noise.[52]

Although we have discussed cyberwarfare in connection only with the US Army, the other service branches are also taking equal parts in US Cyber Command. Each branch of the US military has its own cyberwarfare component. However, the US Army Cyber Command is the first new Army branch since the creation of Special Forces in 1987,[53] and as such it deserved special emphasis in this section.

Figure 3. Soldiers work together during cyberwarfare training in 2011.
(Photo by Staff Sergeant Brian Rodan, US Army)

The evolution of cyberwarfare is proceeding at the speed of an electromagnetic pulse in a fiber optic cable, almost literally. Its critical importance to warfare is recognized, but as with any new military capability the rules of engagement are still evolving. For example, how do you apply "cyber proportionality" to cyberwarfare (with "proportionality" colloquially meaning a tit-for-tat response in warfare)?

Cyberwarfare makes distance and geographic location irrelevant. It also makes detection and deterrence extremely difficult. A cyberattack can start at an instant and without warning. Some experts argue that an all-out cyber assault can potentially do damage that can be only exceeded by nuclear warfare.[54] Unlike a country's nuclear weapons, which a country may reveal in part as deterrence, cyberweapons remain shrouded in secrecy. The secrecy is integral to their usefulness. If a nation is aware of a potential adversary's cyberweapons, they can do more to protect themselves against them.

Cyberwarfare is a large and important area of modern warfare, and there are many books dealing with the subject. However, one aspect that almost all books fail to mention is that cyberwarfare often involves the application of nanoweapons. This raises a question: How does cyberwarfare involve nanoweapons? Let's start with a definition: Nanoweapons are any military technology that exploits the power of nanotechnology.[55] This raises another question: What is nanotechnology? The United States National Nanotechnology Initiative website provides the following definition: "Nanotechnology is science, engineering, and technology conducted at the nanoscale, which is about 1 to 100 nanometers."[56] For those not familiar with the metric system, the diameter of a human hair is about 1000 nanometers. This means that nanotechnology is not visible to the naked eye or even under an optical microscope. Obviously, cyberwarfare involves using high-end computers. Newer high-end computers feature Gen 7 and Gen 8 Intel processors, which include nanoelectronics with 14nm dimensions.[57] When the military uses these computers for cyberwarfare, they are by definition using nanoweapons. This is a reality. To be clear, the processor does not use only nanoelectronics. Only one element of a product needs to have technology with at least one dimension in the nanoscale to qualify as nanotechnology.[58] For example, a guided missile having a guidance system that includes some nanoelectronics is a nanoweapon. In addition to their cyberwarfare thrusts, the US Army is employing numerous nanoweapons, including guided munitions. As I asserted in my book, *Nanoweapons*, the characterization of the US Army as "boots on the ground" may change to "nanoweapons on land."[59]

For our purposes, I chose to highlight that one of the most important roles the US Army is playing in the Third Offset Strategy is its role in cyberwarfare. Now, let's turn our attention to the US Army's strategy for robotics and autonomous weapon systems.

On March 8, 2017, the US Army published its Robotics and Autonomous Systems (RAS) Strategy. In it, the Army said it is setting "realistic objectives in the near-term (2017–2020), feasible objectives in the mid-term (2021–2030), and visionary objectives for the far-term (2031–2040)."[60]

The RAS calls for midterm development of unmanned combat vehicles. The Army asserts it seeks to "introduce unmanned combat vehicles designed to function and maneuver across variable and rough terrain under combat conditions."[61] Although the Army expects the initial robotic or autonomous combat vehicle to "have optionally manned, teleoperated, or semi-autonomous technology."

The deployment of autonomous unmanned combat vehicles would be in violation of DoD Directive 3000.09. However, given advancements in AI, the widespread availability of the technology, and the development of autonomous weapons by China and Russia, my judgment is that the US Army is hedging its bets.

In the near term, the US Army plans to focus on programs that increase situational awareness and lighten carried loads for dismounted forces, improve sustainment via automated ground resupply, improve route clearance systems, and improve explosive ordinance disposal (EOD). Let's consider situational awareness. In its simplest form, situational awareness means knowing what is going on around you. During ground combat, situational awareness can mean the difference between life and death. Situational awareness also refers to the integrated web of networks, servers, storage devices, and analysis and management software that consume data to enable decisions that are more informed at all levels.

One emerging technology is the throwable robot, such as the 1.2 pound, dumbbell-shaped Recon Scout Throwbot,[62] a remote-controlled device that uses infrared optics and relays video. The Recon Scout Throwbot can be tossed into hostile territory, even in the dark, and provide imagery that would otherwise be unattainable.

Getting a complete picture of the combat theater is also critical to making informed decisions. To that end, each branch of the military is developing its own version of the Distributed Common Ground System (DCGS).[63] The DCGS is a network of deployable reconnaissance and surveillance devices that gather, process, and link intelligence, surveillance, and reconnaissance (ISR) data across different databases. The data is stored in situ and uploaded when connectivity becomes available, typically to a network of remote servers, to enable military cloud computing.[64]

The US Army's version of DCGS is DCGS-A, and it brings imagery and intelligence together in a single system.[65] To that extent, it is semiautonomous. A US Army analyst can look at one screen to analyze imagery, signals, human intelligence, and biometrics, allowing for comprehensive situational awareness of the combat theater. The cloud computing aspect of DCGS-A is reducing costs by streamlining intelligence sharing across the military services. It enables the US Army to do a Google-like search to find people and locations. Also like Google it can show locations in a rotatable 3-D diagram.[66]

Obviously, we have only touched on some the US Army's semiautonomous capabilities. In addition to the US Army's cyber capabilities, it deploys thousands of robots, especially to defuse IEDs (i.e., improvised explosive devices). However, these robots are "in-the-loop," meaning they are under human control, including offensive robots such as the wireguided or wireless BGM-71 TOW antitank missile.[67]

US AIR FORCE SEMIAUTONOMOUS WEAPONS

The US Air Force unmanned aerial vehicle (UAV), for example the General Atomics MQ-9 Reaper drone,[68] is an aircraft without a human pilot aboard. In general, Air Force drones are remotely piloted, but most have semiautonomous capabilities built in, including attitude stabilization and hold, hover, and position control, automatic landing and takeoff, automatic return-to-home upon loss of control signal, and GPS navigation, to name a few. This semiautonomy comes from the drones' multiple sensors.[69] However, although the MQ-9 Reaper and other drones feature semiautonomous capabilities, the release of weapons is always under human control.

Drones have proved highly effective in combat and in the war against terror. In addition, drones are cost effective. The cost to build an MQ-9 Reaper drone is approximately $14 million, compared to about $180 million for an F-35 Joint Strike Fighter. The US military favors drones for highly dangerous or laborious missions. In fact, their popularity has led to a severe drone pilot shortage.[70] As a result, the US Air Force is exploring increased levels of autonomy in drones to allow one pilot to manage several at once. To that end, DARPA is looking to build drones that communicate with one another as well as with their operator. This would enable a single pilot to preside over six or more drones simultaneously.[71] However, the release of weapons would still be under human control.

The US Air Force utilizes numerous semiautonomous systems in all weapon systems under their command. Let's take one example to illustrate this point. The Air Force Space Command (AFSPC), a major command of the United States Air Force, oversees the design, acquisition, and operation of most DoD space systems. In general, DoD space systems have numerous semiautonomous capabilities that essentially eliminate human guidance, including switching to backup capabilities, rebooting, and scheduled data transmissions.

Over the last two decades, the US Air Force has developed a family of long-range, semiautonomous stealth cruise missiles called the Joint Air-to-Surface Standoff Missile (JASSM)—for example, the AGM-158 JASSM.[72] These stealthy subsonic cruise missiles are fourteen feet long, including a Teledyne turbojet engine, a small radar cross section, and a thousand-pound conventional warhead. Numerous US aircraft deploy them, from the B-2A Spirit stealth bomber to the F-16 Falcon. The missile's onboard GPS provides navigation from launch to final target approach, up to 230 miles away. At that point, the missile's infrared seeker takes over. The AGM-158 JASSM line dates back to 1995, and the Air Force plans to replace it with the JASSM-ER (extended range), which can travel up to 575 miles to its target due to its larger fuel tank and a more efficient turbofan engine. In addition, the JASSM-ER has been electronically hardened against GPS jamming signals. The AGM-158 JASSM and JASSM-ER share over 70 percent of the same hardware, which reduces production costs. The missiles comprise over-the-horizon "fire and forget" weapons. The phrase "fire and forget" typically means that, once activated, the weapon will autonomously carry out its intended mission. Since a human activates the weapon's release, "fire and forget" weapons are not autonomous. They are semiautonomous. This illustrates the fine line between autonomous and semiautonomous weapons.

The US Air Force seeks three goals in semiautonomous weapons:

1. Standoff Capability—The ability to launch missiles and drones at a distance sufficient to allow attacking personnel to evade defensive fire from the adversary.
2. Decrease Human Cognitive Demand—Assisting human operators in all ways possible to reduce the cognitive demand modern conflict requires. For example, instead of a drone pilot watching a screen

for endless hours, a computer algorithm could watch the screen and signal an alert only when there is a change, such as a truck arriving at a location.

3. Force Multiplication—The US Air Force seeks hardware and artificial intelligence that dramatically increases its effectiveness to accomplish greater military goals at less cost and with a smaller human force. For example, DARPA's program to allow one operator to control six or more drones will ease the demand for drone pilots

THE US COAST GUARD AND US MARINE CORPS SEMIAUTONOMOUS WEAPONS

The US Coast Guard is one of the nation's five military branches.[73] It is part of the Department of Homeland Security. The Coast Guard operates around the world, with the goal of protecting US maritime interests. It targets terrorists, smugglers, illegal aliens, and environmental polluters. It also assists other branches of the military during conflict, and the Coast Guard has often helped rescue people in trouble at sea. The Coast Guard is an armed military force, and its cutters have small arms, pyrotechnics, .50-caliber, 76 mm, or 25 mm machine guns, and shoulder line-throwing guns (i.e., guns that are able to throw a line, such as a wire cable, a long distance).[74] The US Coast Guard operates the Coast Guard Cyber Command (CGCYBERCOM), with the following mission:

> The CGCYBERCOM mission is to identify, protect against, enhance resiliency in the face of, and counter electromagnetic threats to the Coast Guard and maritime interests of the United States, provide cyber capabilities that foster excellence in the execution of Coast Guard operations, support DHS cyber missions, and serve as the Service Component Command to US Cyber Command.[75]

Although the US Marine Corps is a separate and distinct service branch, it has close ties to the US Navy.[76] The Department of the Navy oversees both service branches. However, each service branch has its own autonomous leadership, and both leaders are members of the Joint Chiefs of Staff. In addition, both branches are under the civilian oversight of the

Secretary of the Navy. Obviously, the US Marines benefit from any semi-autonomous weapons developed by the US Navy. However, the US Marine Corps has a unique mission that requires they engage in amphibious warfare. In this regard, the Marine Corps, like the US Army, also engages in testing remote-controlled unmanned ground vehicles, like the Modular Advanced Armed Robotic System, which can perform reconnaissance but can also be equipped with a grenade launcher and a machine gun.[77] The US Marines, like other branches of the US military, engage in cyberwarfare. In October 2009, the Marine Corps established the Marine Forces Cyberspace Command (MARFORCYBER), with a threefold mission (Appendix I), in support of US Cyber Command.[78]

Our overview of the US military's semiautonomous weapons has been representative of the US military's semiautonomous weapons and not comprehensive. The examples in this chapter provide insight into the US military's strategy and deployment of semiautonomous weapons.

The United States is not the only country developing semiautonomous weapons, though. China and Russia have also been working toward developing their own versions.

CHINA'S SEMIAUTONOMOUS WEAPONS

China is the world's second-largest economy, just behind the United States, and has the second-largest defense budget. It is also as a nuclear weapons state.

Unlike the United States, China has no directive banning autonomous weapons, although, in 2016, the country did publish a position paper questioning the adequacy of existing international law to address autonomous weapons, essentially calling on the UN Security Council to address developing new international laws to cover those weapons.[79] However, while China's appears to understand that autonomous weapons present new and unprecedented challenges in warfare, the country is still actively engaged in expanding its AI capability in weapons.

The United States recognizes that the DoD will need to collaborate with companies in the private sector, such as Google and Facebook, whose rapid advances in the AI field will prove critical to the DoD's Third Offset

Strategy. It now appears that China is of the same mind. For example, in 2016, Tencent, developer of WeChat, a Facebook competitor, established an AI research laboratory and began investing in US-based AI companies.[80] In January 2017, Qi Lu, a veteran Microsoft artificial intelligence specialist, left the company to become chief operating officer at Baidu, a giant search firm that rivals Google. Mr. Lu will oversee the company's plan to become the AI global leader.[81] These moves have US military strategists concerned. They question whether the Chinese are merely imitating US AI advances or are engaged in independent innovation that will overtake US AI capabilities.

In August 2016, the state-run newspaper *China Daily* reported:

> China's next-generation cruise missiles will be developed based on a modular design, allowing them to be tailor-made for specific combat situations, and will have a high level of artificial intelligence, according to a senior missile designer.[82]

The Chinese missile appears to be a response to the United States Navy's LRASM (Long Range Anti-Ship Missile), a stealthy, long-range, semiautonomous, anti-ship cruise missile expected to deploy in 2018.[83] The LRASM uses artificial intelligence technology to avoid defenses and make final targeting decisions.

The next-generation Chinese cruise missile typifies a strategy known as "remote warfare," building large fleets of small ships that deploy missiles able to attack an adversary with larger ships, such as aircraft carriers.

In October 2016, a White House report on AI pointed out that China is publishing more on AI research than US scholars, but US military strategists say China has only recently begun to make AI a priority in its military systems.[84] In addition, China has close ties with Silicon Valley companies, in terms of investment and research. As mentioned previously, the open nature of the American AI research community has made the most advanced technology available to China.

In 2016, China brought online the world's fastest supercomputer for its time, the Sunway TaihuLight.[85] China's new supercomputer supplanted its previous Tianhe 2, which had been the world's fastest, powered by Intel's Xeon processors. The processors in Sunway TaihuLight are of a native Chinese design, a strong indication of China's high-tech capabilities.[86] Obviously, while China's new supercomputer has numerous commercial uses, it could also perform military functions, such as breaking the encryption used by adversaries. (Note: During the editing of this book, an important event occurred. On June 8, 2018, the US Department of Energy unveiled Summit,[87] which currently ranks as the world's fastest supercomputer.[88]) In a 2015 edition of a major military journal published about once every fifteen years, China admitted having an army of hackers, although intelligence services around the world had known of the existence of China's cyberwarfare units long before the admission. China terms its cyber units the Strategic Support Force, which analysts estimate at anywhere from 50,000 to 100,000 individuals.[89] It is comprised of the cyber warriors of the Chinese People's Liberation Army (PLA) and others in China's military with responsibility for conducting intelligence, surveillance, and reconnaissance.[90]

Western countries have long accused China of aggressive cyber espionage, including hacks into media outlets such as the *New York Times*, as well as the US electricity grid and several natural gas companies.[91] When the US confronted China in 2008 regarding its cyber espionage, China denied it. In addition to waging cyberwarfare against critical infrastructure and the US military, China targeted commercial technology, most importantly American companies' business strategies.[92]

As US defenses against Chinese cyber espionage became more effective, the combination of declining returns and the prospects of US sanctions led Chinese President Xi Jinping to agree to end Chinese cyber espionage against US commercial companies, then against the United Kingdom's commercial companies, and finally against the other G-20 nations' commercial companies.[93] Since that agreement, China has decreased but not completely eliminated their cyber espionage. There are indications that China is spying on the Russian military industrial complex to learn their military technologies.[94]

Like most nations, China has more experience in cyber espionage than cyberwarfare. Analysts judge Chinese cyberwarfare focuses on supporting conventional military operations and influencing public perceptions and policy in other nations,[95] similar to Russia's apparent meddling into the 2016 US presidential election.[96]

Driven by concern that the US military will intervene in the Asia-

Pacific region, China is pursuing cyber capabilities to corrupt US military logistic information systems and the information links associated with command and control, which are critical to the US military's ability to project force around the world.[97]

In summary, China has a long history in cyber espionage and is highly proficient in its cyber espionage operations, but the nation has less experience with cyberwarfare than do the United States and Russia.

Given China's commercial and military developments, it appears the Chinese intend to match or exceed the United States in artificial intelligence and robotics technologies. Their focus is on cyber espionage, and they appear to be making no concerted effort to match US or Russian capabilities in cyberwarfare.

RUSSIA'S SEMIAUTONOMOUS WEAPONS

Russia seems reluctant to embrace any international law that would limit its ability to field autonomous weapons. As previously discussed, Russia's population is small compared to that of the United States and China. From its perspective, Russia sees autonomous weapons as a way to level the population playing field.

During the Cold War–era, Russia deployed an autonomous missile defense system to protect Moscow. In 1972, the Soviet Union and the United States signed the Anti-Ballistic Missile Treaty, which committed both nations to deploying no more than two antimissile defense systems. However, in 1974, Moscow and Washington agreed to limit themselves to one system. The USSR deployed their system near Moscow. The US installed their complex at the Grand Forks Air Force Base in North Dakota.[98]

Originally, the Soviet Union deployed the A-135 system, equipped with the 53T6 missiles, which were nuclear tipped to ensure destroying enemy ballistic missiles as well as potential enemy missile decoys. However, Russia began upgrading the system in 2000. Their new antimissile defense systems around Moscow is designated the A-235 and includes new missiles that employ a kinetic warhead, not a nuclear warhead.[99] Similar to the US Terminal High Altitude Area Defense (THAAD) system,[100] a kinetic warhead relies on its velocity to inflict damage to incoming ballistic missiles. This suggests that the Russian government has confidence in its new technology and has decided to avoid the possible human losses that would result from the radiation of a nuclear explosion. According to *Russia Beyond The Headlines*:

> Russian Aerospace Forces have tested a missile for their antimissile defense system at a firing range in Kazakhstan. According to an RBTH source in the defense industry, this trial was to test the new short-range warhead (62–620 miles) for the A-235 Nudol systems deployed near Moscow.[101]

This information, from the Russian-owned news source, is publicly announcing they have an operational ballistic missile defense system. However, US experience with the THAAD system indicates that such missile defense systems are of questionable reliability.[102] One possible motive for Russia's public announcement may be deterrence. Obviously, if an adversary believes their missile attack against Russia would fail and retaliation would be certain, that knowledge would deter the adversary from attacking.

Russia is also modernizing its small-arms weapons by using AI to enhance their effectiveness. In July 5, 2017, TASS, Russia's state news media, reported:

> Kalashnikov Group, the producer of the famous AK-74 assault rifle, has developed a fully automated combat module based on neural network technologies that enable it to identify targets and make decisions, Kalashnikov Director for Communications Sofiya Ivanova told TASS.[103]

Apparently, the new combat module is, according to TASS, "fully automated" and consists of a gun connected to a console that analyzes image data to identify targets and make AI decisions over human life and death. Ivanova told TASS that the Kalashnikov Group plans to unveil a range of products based on neural networks, which also suggests autonomous weapons.[104]

Kalashnikov's new combat module isn't the first Russian-made autonomous lethal robot. In 2014, officials with the Russian Strategic Missile Force announced through the Russian news agency Sputnik that they would begin deploying armed sentry robots that could autonomously attack intruders.[105]

Closely following on the heels of the TASS Kalashnikov report, Russia announced plans to develop AI-powered missiles. Russia plans to arm its combat aircraft with cruise missiles incorporating artificial intelligence, in order to be able to make decisions on the altitude, the speed, and the direction of their flight.[106] Russia is also upgrading its Granit class missiles. The Russian Navy's most secret anti-ship cruise missile is the Granit, which includes AI. Due to its speed, maneuverability, and power, the Russians called it the "carrier killer." The Russian Navy plans to replace the Granit with the latest P-800 Oniks supersonic missiles, which are smaller than the Granit and also employ AI. The Russian Defense Ministry plans to deploy them on Project 949A Antey submarines and the Project 1144 Orlan heavy nuclear-powered missile cruisers, which will increase the weapons load of vessels twenty-four to seventy-two missiles. The AI programming allows the missiles to operate as a swarm, communicating with each other to determine the best way to attack the intended target.[107]

Let's conclude by examining Russia's cyber capabilities. Arguably, some military analysts credit Russia with having the most advanced cyber capabilities. Although their meddling in the 2016 presidential election shone a spotlight on some of their capabilities, Russia's cyber capabilities far exceed those witnessed during the election. For example, in 2015, Russia attacked the Ukrainian power grid, via a coordinated cyberattack against multiple energy companies in the Ukraine. A quarter of a million people were without power for several hours, which some analysts suggest was a "dry run" for an attack against US power grids. Since modern society is dependent on computers, cyberattacks can cause real-world impacts.[108]

A power grid attack on the United States might not be as limited as the attack on Ukraine's power grid. Ukraine's system operators were able to revert to manual controls to restore power. Given the level of automation used in the United States, that might not be possible. Clearly, our adversaries know that the US power grid is antiquated and fragile. As Ted Koppel, speaking of Iran rather than Russia, pointed out in his bestseller *Lights Out*:

> Iran surely understands that it cannot hope to wage a nuclear war with the United States and win, but Iran will continue pursuing its strategic interests by other means: terrorism, the use of surrogates, and, increasingly, cyber warfare.[109]

In 2015, Lloyd's of London issued a report analyzing what would result from even a limited attack on the US power grid.[110] In it, Lloyd's suggests that an attack on US Northeast's power grid could result in some areas experiencing a weeks-long blackout, resulting in economic losses ranging from $250 billion to $1 trillion. Lloyd's also speculated that a power grid loss could result in breakdowns in logistics systems causing widespread shortages, looting, and rioting. In such a scenario, Lloyd's forecasts the necessity to impose martial law.

What does the future hold? To quote the Center for Naval Analyses report, two things are clear:

1. In keeping with traditional Soviet notions of battling constant threats from abroad and within, Moscow perceives the struggle within "information space" to be more or less constant and unending. This suggests that the Kremlin will have a relatively low bar for employing cyber in ways that US decision makers are likely to view as offensive and escalatory in nature.

2. Offensive cyber is playing a greater role in conventional Russian military operations and may potentially play a role in the future in Russia's strategic deterrence framework. Although the Russian military has been slow to embrace cyber for both structural and doctrinal reasons, the Kremlin has signaled that it intends to bolster the offensive as well as the defensive cyber capabilities of its armed forces. During the contingencies in Georgia and Ukraine, Russia appeared to employ cyber as a conventional force enabler.[111]

In essence, Russia sees cyberwarfare as "constant and unending." Based on all the publicly available information, expect any US conflict with Russia to provoke an offensive cyberattack, among other offensive attacks.

SIGNIFICANT INSIGHTS

Currently, nations like the United States, Russia, and China are deploying weapon systems that require minimal human guidance to carry out their missions, from drones to missile defense systems to sentry robots. Russia, in particular, appears to be including more autonomy in their weapon developments. The next logical advance will be weapon systems that require no human guidance, termed "lethal autonomous weapons" (LAWS). Many see this as the "third revolution in warfare," after gunpowder and nuclear arms. Because of AI's rapid advances in recent years, specifically in the field of machine learning, AI systems are able to recognize complex or subtle patterns in large quantities of data, which enables the system to perform operations equal to or better than humans. Because AI offers significant operational advantages, including replacing human soldiers in harm's way, nations like the United States, Russia, and China are seeking to develop and deploy LAWS. While the United States and China appear to want to draw a line just short of autonomous weapons (i.e., humans out-of-the-loop), Russia does not. In reality, the complexities of the twenty-first-century battlefield may necessitate the deployment of autonomous weapons. The military community no longer debates about whether to build autonomous weapons. The debate now centers on how much independence to give them. This is something the US military has dubbed the "Terminator Conundrum."

Computers are like Old Testament gods; lots of rules and no mercy.

—Joseph Campbell, *The Power of Myth*, 1988

CHAPTER 4

THE NEW REALITY

The development of weapons does not occur in a vacuum. Weapons development is generally the result of perceived threats. When the United States or any nation perceives a threat, it seeks a way to neutralize it. This is what led to the United States' Offset Strategies. Each Offset Strategy addressed a specific well-articulated threat that the United States perceived from potential adversaries. Other nations go through a similar process to develop weapons. The result of this process is a new reality. It may be a period characterized by a large-theater conflict, such as World War I and II, or it may lead to an array of lesser conflicts.[1] These lesser conflicts run the spectrum from limited conventional wars, such as the United States' involvement in Operation Desert Storm in 1991, to ambiguous wars, such as Russia's seizure of Crimea in 2014 using troops with uniforms lacking a national insignia.

In the new reality, there is no peace. Although we may fight a war and win from a military standpoint, the result is that we will still find ourselves engaged in a form of conflict. For example, the period following the end of World War II was a Cold War, characterized by a nuclear arms race between the United States and the former Soviet Union. Currently, after winning the war in Afghanistan in 2001 and the war in Iraq in 2003, the United States finds itself engaged in fighting counterinsurgency and attempting to stabilize each nation.

To articulate the new reality, we must understand 1) the Political Reality, 2) the Technological Reality, and 3) the Weapons Reality. Taken as a whole, they will provide an understanding of the driving forces that underpin the development of new weapons.

THE POLITICAL REALITY

Understand that this view of the political reality is only a snapshot in time, and it has likely changed by the time you are reading this book. I say "likely" because events are unfolding quickly on the world stage. However, understanding the state of events concurrent with writing this book will illustrate how politics shapes weapon developments. Consider each of the elements we discuss in this section a snapshot. When we view the snapshots as a collage, a clear picture of the politics shaping the new reality will be apparent.

In chapter 3, we mentioned that the United States is facing the "Terminator Conundrum." On the one hand, the Defense Department will invest $12 billion to $15 billion in its fiscal year 2017 budget on its Third Offset Strategy. The Third Offset Strategy calls for developing and deploying weapons exploiting artificial intelligence, robotics, and nanoweapons. The US military believes the United States holds a leadership in these technologies, which will translate into military superiority relative to its most capable potential adversaries, China and Russia. On the other hand, the United States operates under DoD Directive 3000.09, which clearly prohibits the US military from deploying autonomous weapons. To be clear, a weapon may be autonomous but DoD Directive 3000.09 requires its deployment be semiautonomous, meaning a human must be in- or on-the-loop. The Defense Department is capable of designing autonomous robotic fighter jets, missiles that can decide when and what to attack, and ships that can hunt and destroy enemy submarines. The conundrum is: Should the US military build and deploy these new autonomous weapons of war?

I believe that we see this as a conundrum because we are asking the wrong question. Our most capable adversaries, China and Russia, are developing autonomous weapons. To my mind, the question becomes: Will the United States be able to maintain military superiority if our most capable adversaries have autonomous weapons that outperform human controlled weapons?

In the short term, and by short term I mean within a decade, we may be able to maintain military superiority. However, given the advancements in artificial intelligence, robotics, and nanoweapons technology that are likely to occur before that time, I question if we can maintain long-term military superiority against capable adversaries like China and Russia, who are not constrained by Directive 3000.09.

This is an ethically and politically charged issue. The use of lethal autonomous weapons (LAWS) generally conjures the concept of "killer robots" roaming the Earth indiscriminately killing combatants and non-combatants. In 2013, a group of nongovernmental organizations who sought to preemptively ban LAWS formed the Campaign to Stop Killer Robots.[2] In July 2015, over 1,000 AI experts, along with 15,000 other endorsers, signed an open letter (Appendix II) warning the world of the threat of an arms race in military artificial intelligence and calling for a ban on autonomous weapons.[3] The letter was presented in Buenos Aires at the Twenty-Fourth International Joint Conference on Artificial Intelligence. Signatories include Stephen Hawking, Elon Musk, and Steve Wozniak, among others.

I mention the Campaign to Stop Killer Robots and the open letter, "Autonomous Weapons: An Open Letter from AI & Robotics Researchers," because they form one element of the political reality and also make a point that there is significant concern among people throughout the world regarding the development of autonomous weapons. Therefore, consider this as the first snapshot. Remember, we are going to view each political event and consider it a snapshot. Later we will view the snapshots as a collage to form a complete picture of the politics forming the new reality.

When the United States announced its Third Offset Strategy, it sparked a new arms race. As previously discussed in chapter 3, AI technology is central to the Department of Defense's Third Offset Strategy. As a result, China and Russia are rapidly moving to incorporate AI into their weapon developments. Russia, for example, has an overt military strategy to build autonomous weapons. As with any Cold War, tensions between the United States, China, and Russia are high. Each is eager to have AI enabled weapons that would deter war or win it if necessary. Consider this the second snapshot.

This next snapshot is a current reality as I write this book. Therefore, consider this reality in its historical context. Although I will specifically discuss North Korea, the snapshot applies to any rogue state that threatens world peace.

North Korea, in violation of UN resolutions to prevent the proliferation of nuclear weapons, continues its relentless pursuit to develop ballistic missiles and nuclear weapons capable of attacking the United States.[4] China, which is North Korea's largest ally and trading partner, appears unable to get North Korea to abandon its nuclear ambitions. This situation is becoming more perilous by the day. Because of threats made by North Korea against the United States, the United States has deployed three carrier groups to the Asia-Pacific region, including the USS *Nimitz*, USS *Carl Vinson*, and USS *Ronald Reagan*.[5] In a side note, only the *Ronald Reagan* is home-ported at Yokosuka, Japan, as part of the US Seventh Fleet. A typical carrier group is composed of approximately 7,500 personnel, an aircraft carrier, at least one cruiser, two destroyers or frigates, and a carrier air wing of sixty-five to seventy aircraft. The United States has also announced the deployment of two submarines to the Asia-Pacific,[6] an Ohio-class guided missile submarine, the USS *Michigan*, and a Los-Angeles-class attack submarine, the USS *Cheyenne*. This armada has more destructive capability than the United States possessed during all of World War II. Lastly, the United States deployed its Terminal High Altitude Area Defense (THAAD) system in South Korea, an antiballistic missile defense system the US military claims will destroy an adversary's intercontinental ballistic missiles during conflict. This has increased tensions between the United States and China, which also sees THAAD as a threat to its own ballistic missile defenses.[7] Consider this the third snapshot.

In addition, China is claiming sovereignty over the South China Sea.[8] To that end, China has constructed seven artificial islands in the South China Sea since 2014. China has turned its new islands into military installations, complete with airstrips, radar arrays, and other assets that enable China to project force in the region. Although Vietnam, the Philippines, Malaysia, and China all claim portions of the South China Sea, China's claim is extraordinary in that it covers most of the sea, extending down to the coast of Malaysia. This prompted the Philippines to bring a suit against China to the Permanent Court of Arbitration in The Hague, which on July 12, 2016, ruled against China.[9] However, China refuses to acknowledge the decision and the court's jurisdiction. The United States asserts that the

South China Sea is an open waterway, and backs this assertion by periodically sending US warships through the region. You may wonder: Why is the South China Sea so important that the United States appears willing to defend it? There are three reasons:

1. According to the Council on Foreign Relations, more than $5.3 trillion worth of shipping travels through the sea each year, which includes $1.2 trillion of US trade.
2. The US Energy Information Administration (EIA) estimates that the China Sea harbors beneath its ocean floor 11 billion barrels of oil and 190 trillion cubic feet of natural gas.
3. 12 percent of the world's global fishing catch comes from the South China Sea and fishing is a crucial industry for China, which is the largest fish producer and exporter in the world. In 2015, China's fish exports were nearly $20 billion (US).

Consider China's sovereignty claim over the South China Sea our fourth snapshot. Again, be mindful that this snapshot may change. Therefore, view it in its historical context.

Lastly, let's examine the political realities associated with Russia. The single largest political issue with Russia is its increasingly expansionistic thrusts.[10] Here are just a few:

- The annexation of Crimea.
- The naming of Ukraine's eastern regions as Novorossiya (New Russia).
- The sponsorship of terrorism and separatism in Ukraine.

Russian expansionistic objectives aim to undermine the United States. These objectives directly oppose the spread of democracy, especially in the former Soviet republics. Russian leadership views Ukraine as part of its "original" territory and considers Ukrainians Russians. Currently, the Ukraine is in a war between the post-revolutionary Ukrainian government and pro-Russian insurgents, supported and assisted by the Russian military.[11] Given their brutal expansionistic thrusts, Russia appears to be Europe's greatest threat. Consider this the fifth snapshot.

Of course, there are numerous other factors, such as the war on terror and the war in Syria, that affect the new reality. However, in my judgment, the five snapshots above, when viewed as a collage, are sufficient to give us a clear picture of the politics that form the new reality. Here is a short summary:

- People throughout the world seek to ban the development and deployment of autonomous weapons.
- Indifferent to the world outcry to ban autonomous weapons, there is a new Cold War raging between the United States, China, and Russia to develop artificially intelligent weapons, and each side sees their military strength as being dependent on the effectiveness of such weapons. For now, each nation is deploying mostly semiautonomous weapons. As AI technology rapidly matures, autonomous weapons will almost certainly find their way to the battlefield.
- While the US has a self-imposed moratorium on the development of autonomous weapons, China and Russia do not.
- The world is a tinderbox of tensions. North Korea, as of this writing, is pursuing a course that could ignite a nuclear war. In addition, China's claim of sovereignty over the South China Sea and Russia's expansionism objectives could also ignite a conflict.

Before we leave this section, let us address an important question. Would it be possible to establish a treaty to ban autonomous weapons? Unfortunately, I doubt it. There is too much suspicion and tension. It would be difficult to get to the point where all sides could agree. Russia, in particular, has a smaller population than the United States and China. As such, they see autonomous weapons as a way to make the population playing field equal. Based on their past statements, it appears the United States and China would be receptive to banning autonomous weapons, but unless Russia accedes the treaty would be doomed to failure.

While it's true that the nations of the world have come together to ban certain types of weapons, such as chemical and biological weapons, space-based nuclear weapons, and blinding laser weapons, those weapon classes are different from autonomous weapons. Let's consider some examples that highlight the differences.

Although biological weapons have the potential to be weapons of mass destruction, their control during conflict would be a major issue for every nation. There is no way to ensure a biological weapon would only affect the adversary. Once released, it could spread beyond the adversary's borders and begin killing indiscriminately. This aspect of biological weapons worries all nations and made them a perfect candidate for nations to agree to ban. Thus, the Biological Weapons Convention (BWC) became the first multilateral disarmament treaty banning the development, production, and stockpiling of such biological weapons, and it took effect on March 26, 1975.[12]

Most nations agree that chemical weapons offer no sustainable strategic advantage. Even when chemical weapons made their debut in World War I, nations were able to quickly develop countermeasures against them, such as gas masks. As chemical weapons technology advanced, so did the countermeasures. Given the low-tech nature of chemical weapons technology, in time all nations would have them. Most importantly, chemical weapons, like biological weapons, are hard to control. For example, they are subject to changing weather condition. Hence, given their relative ineffectiveness and inherent difficulty to control, chemical weapons found no proponents. Thus, the Chemical Weapons Convention (CWC), a multilateral treaty that bans chemical weapons and requires their destruction, became effective in 1997.[13]

As is evident from the above, the ban on chemical and biological weapons during warfare has a strong support base. That support, especially for biological weapons, arises due to the difficulty of controlling them. In the case of chemical weapons, the support stems from their strategic ineffectiveness.

The ban on space-based nuclear weapons, or as it is formally termed, "Treaty on Principles Governing the Activities of States in the Exploration and Use of Outer Space, including the Moon and Other Celestial Bodies," forms the basis of international space law.[14] The United States, the United Kingdom, and the Soviet Union opened the treaty for signature on January 27, 1967. It became effective on October 10, 1967, and, as of July 2017, 107 countries are parties to the treaty. In addition, 23 have signed the treaty but have not completed ratification, including North Korea. Although this treaty has held, even among nonsignatories, the United States, China, and Russia all have the capability to launch a space-based nuclear weapon attacks, such as an EMP (i.e. electromagnetic pulse) attack. An EMP attack involves detonating a nuclear weapon in the space above an area to unleash an electromagnetic pulse that would disrupt and destroy electronics within that area. The EMP attack can focus on only part of a nation or can include the entire nation, depending on the type of attack. No nation has implemented an EMP attack because of the nuclear retaliation such an attack would provoke. Unfortunately, even a small, less technically advanced nation, like North Korea, could also launch such an attack. As I write, North Korea has two satellites circulating the Earth, and no one but North Korea knows their purpose. They pass over the United States daily. Each pass puts them over a different portion of the United States. Their size suggests that each could contain a small nuclear weapon. Their altitude is three hundred miles, which experts consider perfect for an EMP attack. I am mentioning this to make a point. The ban on space-based nuclear weapons may be holding, but it may easily be broken.

On October 13, 1995, the United Nations issued "The Protocol on Blinding Laser Weapons," which came into force on July 30, 1998.[15] The thinking behind banning these weapons was that they caused superfluous and unnecessary suffering. However, as of the end of April 2016, only 107 states had agreed to the protocol. This is slightly more than half of the 193 member states of the United Nations. In addition, both the US military and private companies are making laser weapons intended to cause temporary blindness, termed "dazzlers."[16]

My point in discussing these treaties, which have banned certain classes of weapons, is that a strong case exists that:

- Controlling those weapons is difficult to impossible, such as in the case of biological and chemical weapons.
- The treaty is easily broken or sidestepped, such as in the case of space-based nuclear weapons or blinding laser weapons.

Now, if we consider a treaty that seeks to ban autonomous weapons, what makes them different from the above banned weapons?

1. The countries fielding autonomous weapons express no issues related to controlling them. For example, Russia is currently fielding

autonomous sentry robots at its Moscow antimissile defense installation. In addition, Russia is asserting it will develop and deploy more autonomous weapons. The United States exempts its cyber-warfare weapons from its assertion to only develop and deploy semi-autonomous weapons. This is likely due to the time-critical nature of cyberattacks. From these examples, we can infer that the major players feel they can control autonomous weapons.

2. Treaties governing autonomous weapons may be easily broken or sidestepped. The United States, for example, is developing humans on-the-loop semiautonomous weapons. This means the weapon acts autonomously, but a human can monitor its actions and shut it down. The operative phase is "the weapon acts autonomously." Factually, with this caveat (i.e., humans on-the-loop), the United States is building autonomous weapons. This means a treaty banning autonomous weapons could be broken with a flip of a switch.

3. Lastly, it's not clear we can enforce a treaty that bans autonomous weapons. Autonomous weapons can be manufactured with off-the-shelf AI components, which makes treaties banning them difficult to enforce. By comparison, nuclear weapon treaties are easier to enforce since nuclear weapon manufacture is difficult to hide. Yet, even nuclear weapon treaties are difficult to enforce. For example, let's consider the Intermediate-Range Nuclear Forces (INF) Treaty. Ronald Reagan and Mikhail Gorbachev signed the INF Treaty in December 1987. It banned ground-launched ballistic and cruise missiles with ranges of 300 to 3,400 miles.[17] Both countries appeared to be abiding by the treaty until Russia began testing cruise missiles in 2008. The Obama administration concluded in 2011 that Russia's cruise missile tests were a compliance concern. In May 2013, Rose Gottemoeller, the State Department's senior arms control official, confronted Russian officials regarding their possible treaty violation, which apparently had no effect. Russia conducted further tests and deployed INFs through 2017, which the United States claims are clear treaty violations.[18] Russia does not deny its violations but argues that nearly all its neighbors are developing these kinds of weapon systems. Given Russia's enormous strategic nuclear missile capabilities, it's not clear they need

INFs. However, my point is simple. Since it is obviously difficult to detect and stop violations to nuclear weapons treaties, would it even be possible to detect and stop violations to an autonomous weapon treaty? You be the judge.

THE TECHNOLOGICAL REALITY

The brain of semiautonomous and autonomous weapons is artificial intelligence. Therefore, to understand the technological reality we need to understand the current state of artificial intelligence technology and its development direction.

Although the United States, China, and Russia all develop and deploy semiautonomous and autonomous weapons, the artificial intelligence within the weapons remains relatively crude. How can I make such an assertion?

All current semiautonomous and autonomous weapons depend on smart agents to perform their mission. As we discussed in chapter 2, smart agents are computer algorithms that perform intelligent behavior with minimal human intervention and are even able to learn from experience.

While smart agents can perform some amazing feats that appear to mimic human intelligence, they do not embody human intelligence. The way most smart agents work is though programs that define specific rules and patterns that allow them to perform their missions. For example, if your smartphone has a chess-playing application, the brain behind it is a smart agent. It may have numerous rules and a database that includes numerous games of chess masters. It searches the database for a pattern similar to the chess move you may have just made and uses its rules to determine what it should do, based on the probability of various outcomes. My point is, the smart agent is not thinking in a human sense. It is analyzing chess moves and following rules that suggest a high probability its moves will eventually lead to your defeat. Unless you are a master chess player, the probability is that this smart agent will win. Because it appears to mimic human thought, we assume it is thinking. It is not. It is analyzing patterns and making calculations. If it were truly able to think, it should be able to play a simpler game like checkers, but it cannot. The reason it cannot is that it has no programming to play checkers. Even if the smart agent learns from experience, it typically learns in only one specific area, like chess.

If we extend this example to semiautonomous and autonomous weapons, they are subject to the same limitations. They can mimic human thinking and even perform specific functions faster than humans, but they are not thinking. They are following their programming.

How do computers differ from human brains? The answer to this question depends on what type of computer you are talking about and how it's programmed. Simple computers that use decision-tree programming and databases to perform functions, which is typically how smart agents work, are as close to human brains as birds are to jet planes. While they both fly, their method to accomplish flight is vastly different. How does the average human brain work that makes it so different?

The average brain has about 100 billion neurons, which are specialized cells capable of transmitting electrical or chemical signals through connections termed synapses to other neurons. Said simply, synapses connect neurons. Each neuron connects to about 10,000 other neurons. If you do the math, this means the human mind has between 100 trillion and 1,000 trillion synaptic connections. Neurons process all information that flows into, out of, and within the central nervous system. In addition, and to oversimplify, the configuration enabled through the synaptic connections allows the brain to store information or, as we often describe it, form memories.[19] Mathematical calculations of the number of configurations enabled through the synaptic connections suggest the human brain's memory capacity could be as high as 1,000 terabytes (i.e., a unit of information equal to one million million) of data. By comparison, the 19 million volumes in the US Library of Congress are about equal to 10 terabytes of data. In addition to neurons, 90 percent of the brain is composed of glial cells, which surround neurons, provide nutrients to neurons, and act as insulation between neurons.

Until recently, neuroscientists focused their attention on the brain's 100 billion nerve cells called neurons because neurons appeared to communicate with each other via synapses. By comparison, neuroscientists believed that glial cells simply supported neurons. Recent studies, however, suggest glial cells are vital to the development and function of synapses.[20] Through advances in imaging technology, neuroscientists learned that glial cells were actually communicating by chemical means, among themselves and with neurons. This has led to an important conclusion: glial cells play a vital role in brain cell communication, and perhaps in the development of human intelligence. When Albert Einstein died in 1955, scientists removed his brain from his body and preserved it in a jar of formaldehyde. Over the next thirty years, scientists examined portions of his brain, hoping to discover clues to Einstein's genius. They found that Einstein's brain was average in size and normal in the number and size of its neurons. However, in the late 1980s, scientists discovered that Einstein's brain contained more glial cells than normal, especially in the association cortex, an area of the brain involved with imagination and complex thinking. While this is just anecdotal information, it does suggest that glial cells may be critical to intelligence.

We are still learning how the human brain works. However, we do know that its memory capacity likely exceeds all the information stored in the Library of Congress. We also know that each neuron connects with up to 10,000 other neurons, passing signals to each other through as many as 1,000 trillion synaptic connections. Some computer scientists estimate that this is equivalent to a computer with a one trillion bit per second processor.[21]

Based on our understanding of the human brain, we've begun placing more emphasis on building neural networking computers. The reason is simple. Rather than programming computers for endless hours, it is easier and faster to let them learn on their own. In fact, some experts believe that artificial neural networks (ANN) that learn about the world via trial and error, similar to the way toddlers learn, might develop something akin to human learning. It turns out that this approach was correct. The method not only works, but it works faster and better than the traditional methods of programming. As an example, let's consider Google Translate.

Google introduced its language translation application in 2006. Currently, about 500 million monthly users translate 140 billion words per day, from one language to another. In 2006, Google created Translate by writing a comprehensive program that laid out the rules of logical reasoning, plus knowledge about the world. To translate from English to Japanese, for example, they programmed all of the grammatical rules of English, plus all of the definitions contained in the *Oxford English Dictionary*. Next, Google programmed all of the grammatical rules of Japanese, plus the words in the Japanese dictionary. Using this method, Google sought to translate sentences from one language to another. Unfortunately, performing translation using this approach often loses meaning. For example, "minister of agriculture" may translate as "priest of farming." In addition, the approach was time consuming. Languages tend to have as many exceptions as rules. The approach works extremely well when rules and definitions are clear, such as with chess. Smart agents that use this approach are generally superior chess players versus their human opponents. This is why many AI researchers believed the road to human-level intelligence was paved with smart agents. It turns out they are wrong. The road to human-level intelligence is likely paved with neural network algorithms.

In the case of Google Translate, Jeff Dean, Andrew Ng, Greg Corrado, and Quoc Le formed what Google refers to as Google Brain in 2011.[22] Their thrust was to forward the use of neural network computing to solve some of the great challenges facing AI. It turns out that as sophisticated as smart agents are, they cannot perform simple tasks a toddle could do, such as identifying a cat. The Google Brain group realized that human toddlers didn't learn by memorizing rules and dictionaries. They learned from experience. The Brain group reasoned that artificial neural networks (ANNs) might be able to do something similar and learn the way toddlers learn, essentially bypassing the need for programming. This would mean ANNs would be capable of modifying its actions based on the information it receives (i.e., feedback) rather than by rigidly stepping through a fixed array of operations.

In its first year, Brain's experiments in the development of a neural networking machine began to emulate the capabilities of a one-year-old toddler. Google's speech-recognition team swapped out part of their old system for a neural network. This resulted in the best translation quality improvements in twenty years. It became increasingly obvious to Google that the path to a machine understanding meaning was through neural networking. Now, similar to the US military, Google sees AI as critical to compete against the likes of Facebook, Microsoft, Apple, Amazon, and others around the world.

Neural network computing is surfacing as critical to enabling machines to think, as opposed to following programmed instructions. This is a major paradigm shift, and it seems to be gaining traction.

The second big paradigm shift has to do with Moore's law. According to Marc Andreessen, American entrepreneur, investor, software engineer, and cofounder of Netscape, Moore's law has flipped.[23] If you recall from chapter 1, Moore's law asserts that the number of transistors in a dense integrated circuit doubles approximately every two years, while the price of the integrated circuit remains the same. Andreessen asserts that new integrated circuits are coming out equivalent to their predecessors, but at half the cost. This means that the cost of computing is dropping. This will accelerate the "Internet of Things" we discussed in chapter 2, allowing a new, highly interconnected world as never before. In addition, low-cost processors can enable parallel and distributed systems to cost effectively solve problems that were unthinkable even a few years ago. I like to think of this as a corollary (i.e., a result in which the proof relies heavily on a given theorem) of Moore's law. My reasoning is that if Moore's law asserts that we can get an integrated circuit with twice the density of circuits at the same price every two years, it seems to follow we should be able to get an integrated circuit with the same density at approximately half the price every two years. I do not see any contradiction between the two assertions. Said more positively, both can be true, and to my mind they are. In fact, as technology advances, products improve and prices drop. As the US Bureau of Labor Statistics indicates, prices have dropped dramatically in almost every tech sector for the last eighteen years, especially in computer hardware. Let's refer to this as **the Law of Decreasing Cost Returns**. The idea is that as technology increases the cost of former generations of that technology decrease. For example, a television you buy today will cost less than one purchased two years prior, even though it is approximately the same in size and capability. The price drop is due to decreases in component costs and increases in manufacturing efficiency. Similar to Moore's law, this is an observation of a trend and not a physical law.

As we mentioned previously, Moore's law can more generally be stated as the Law of Accelerating Returns. This restatement suggests it is an observation regarding the accelerated pace of human innovation in any well-funded technology field, such as solid-state electronics. Therefore, as Moore's law is likely to end by approximately 2025 as it strictly applies to integrated circuit lithography, the Law of Accelerating Returns implies solid-state technology is likely to continue its accelerated improvement.[24]

The third and last paradigm shift has to do with data. Many sources predict data growth is becoming exponential as we approach the year 2020 and broadly agree that the size of the digital universe will double every two years, which translates to a fifty-fold growth from 2010 to 2020.[25] This is because most data is moving online and becoming mobile via smartphones, and sensors are recording almost everything. In essence, data is fueling a digital revolution. Organizations that have access to the data pertinent to their business and the ability to process that data to glean important insights will have a competitive edge. In addition, growth of data in digital form is a language that machines can understand. Neural network algorithms enable machines to accelerate deep learning from that data.

Now let us address one of the most critical questions: When, expressed as a point in time, will we have a computer that equals a human brain? If you do a Google search to address this question, you will find answers all over the map. Most people address this question by extrapolating Moore's law to determine when we will have a computer with an equivalent number of artificial neural networks as the human brain's natural neural networks. However, this does not mean the computer has human-level intelligence. It means the computer may have the processing power of a human brain. I used the phrase "may have" because we do not know if an artificial neuron is equivalent to a human brain's neuron. In addition, the human brain is not strictly a digital computer. While it is true that neurons either fire or don't fire, the neuron signals travel through the neuron via biochemical pathways, which perform a type of signal processing. Where does this leave us? Numerous people, from computer scientists to neuroscientists, have been trying to address this question. Ray Kurzweil,[26] one of the leading futurists in artificial intelligence, and Chris F. Westbury,[27] a research psychologist, estimate the processing speed of a human brain to be approximately 20 petaflops.[28] (A petaflop refers to computing speed equal to one thousand million million operations per second.) Other researchers argue that it is closer to 40 petaflops.[29] If we want to be conservative, we can double the highest estimate and assert the human brain has a processing capability of about 80 petaflops. While this is enormous, the Chinese have built a supercomputer, Sunway TaihuLight that has a processing power estimated at 93 petaflops.[30] However, it is not equivalent to a human brain. It lacks the programming that would allow it to equal human-level intelligence. However, let's assume the Sunway TaihuLight was able to learn via neural networks. If it learned at about the same rate as humans learn, it could exhibit human-level intelligence, equivalent to an adult, in a matter of decades. My point is that software, as well as hardware, plays a critical role when it comes to achieving a machine with human-level intelligence.

Key AI Projections

Because of the enormous strides in computer technology, evidenced the US Department of Energy supercomputer Summit, and the development of self-learning artificial neural networks, evidenced by Google Brain, I project the emergence of human-level intelligence on or before 2040. This means AI that would be able to pass the Turing Test (i.e., a conversation between a human adult and an intelligent machine would be indistinguishable to an objective third party, and that third party would find it impossible to tell the human and the machine apart). This is also consistent with the projected emergence of human-level intelligence in the 2040–2050 timeframe (at a 50 percent probability) that Vincent C. Müller and Nick Bostrom provide in their "Future Progress in Artificial Intelligence: A Survey of Expert Opinion."[31] I mention their survey because the projections of AI researchers regarding when a computer will equate to human-level intelligence has ranged from "never" to "twenty years from now." The "now" represents whenever you ask the question. Müller and Bostrom developed a survey to get a more accurate projection. In addition, their survey asserts that we should expect once a system reaches human-level intelligence, that the "systems will move on to superintelligence in less than 30 years thereafter." Müller and Bostrom define superintelligence as "any intellect that greatly exceeds the cognitive performance of humans in virtually all domains of interest."[32] Kurzweil and others define this point in time as a singularity.[33] Therefore, if we use my projection regarding the emergence of human intelligence (i.e., 2040), and Müller and Bostrom's projection that superintelligence will emerge within thirty years after the emergence of human intelligence, the singularity will occur on or around 2070. Please note, my projection regarding the emergence of human-level intelligence favors Müller and Bostrom's more optimistic projection.

Summary of the New Technological Reality

Let us summarize the new technological reality, based on previous discussions:

- Neural networking is enabling computers to learn to perform functions from experience, without being programmed to do so.
- Moore's law is flipping, making computer processors significantly less costly about every two years. We termed this the Law of Decreasing Cost Returns. This suggests that "Internet of Things" trend we discussed in chapter 2 will accelerate, and eventually almost everything will connect to the internet. This does not contradict Moore's law. It as a corollary to Moore's law.
- Data is exponentially doubling every two years, which in turn will enable computers with neural network algorithms to learn numerous new and difficult tasks, without being programmed to do so.
- Supercomputers, equivalent to the human brain's processing power, are emerging now and will be common in the coming decades.
- Neural network computing suggests that on or around 2040, supercomputers will reach human-level intelligence, which is consistent with Müller and Bostrom's most optimistic projection.
- Superintelligent machines will emerge by or in fourth quarter of the twenty-first century, which is consistent with Müller and Bostrom's projection, assuming supercomputers reach human-level intelligence on or around 2040–2050.

Before we leave this section, I should mention that projections regarding when a machine will reach human-level intelligence or when the singularity will occur differ significantly between various researchers. As I mentioned, Müller and Bostrom attempted to get a more accurate projection via their survey. Rather than being argumentative, I will simply say that my projections favor their more optimistic projections reported in their survey. However, I felt that I needed to provide my projections and the methodology used to reach them. With this information, you can use your own judgment regarding the technological reality.

THE WEAPONS REALITY

Consistent with DoD Directive 3000.09, the United States asserts it does not deploy autonomous weapons. However, we need to carefully understand DoD Directive 3000.09 because current US development and deployment of semiautonomous weapons provides an ample foundation to field autonomous weapons.

DoD Directive 3000.09 asserts:

> Autonomous and semiautonomous weapon systems shall be designed to allow commanders and operators to exercise appropriate levels of human judgment over the use of force.[34]

Notice two points: 1) the directive mentions both autonomous and semiautonomous weapon systems, and 2) the directive requires the design of such weapons to allow "appropriate levels of human judgment over the use of force."

Clearly, DoD Directive 3000.09 does not ban autonomous weapons but seeks to ensure "appropriate levels of human judgment over the use of force." In chapter 3, we discussed three levels of human control over semiautonomous and autonomous weapons, namely, 1) humans in-the-loop, 2) humans on-the-loop, and 3) humans out-of-the-loop. The second category, humans on-the-loop, means that autonomous weapons may be designed and deployed, but humans must be able to observe their operation and have the ability to shut them down. This means the weapon can act autonomously unless a human commander overrides the autonomy. Under this premise, the Department of Defense is developing:

> robotic fighter jets that would fly into combat alongside manned aircraft. It has tested missiles that can decide what to attack, and it has built ships that can hunt for enemy submarines, stalking those it finds over thousands of miles, without any help from humans.[35]

My point is that the distinction between semiautonomous and autonomous weapons is an extremely fine line. Take a simple example: Let's consider the Aegis Combat System used by the Navy. If a carrier group comes under attack from numerous sources, the Aegis system is able to track all threats and direct missiles to neutralize the threats. This occurs with an operator and commander on-the-loop. Given the threat to the carrier group and the chaos of the battle situation, do you think it's likely the commander will override any decisions made by the Aegis system? Because of the time-critical aspects regarding the defenses, I suspect the commander can do little but oversee the system's performance. For all practical purposes, the Aegis Combat System will be acting autonomously.

Our most advanced adversaries, China and Russia, have not imposed a directive similar to 3000.09 on themselves. Russia, in particular, is making its intentions to deploy autonomous weapons known. China, while not as overtly asserting its position to deploy autonomous weapons, seems to have little objection to deploying them. In fact, China plans on deploying Russia's S-400 antimissile defense system, widely consider among the best antimissile defense system in the world.[36] The S-400's capabilities include multifunction radar, autonomous detection and targeting systems, and anti-aircraft missile systems.[37] While we cannot ascertain the level of autonomy, it clearly includes some autonomous functionality.

Currently, all semiautonomous and autonomous weapons employ smart agents. In other words, they fall short of human-level intelligence. However, given the exponential progress of AI technology, we should expect our most advanced adversaries to deploy autonomous weapons with human-level autonomy in roughly the 2040–2050 timeframe. This is fully consistent with the progress of AI technology achieving human-level capabilities in that timeframe. When this occurs, the capabilities of autonomous drones will exceed our best-piloted fighter aircraft. The reasoning behind this assertion is simple. Without needing to concern itself with a pilot's safety, autonomous drones will be able to execute high-g (high gravitational forces induced by acceleration) maneuvers. In general, a human pilot, equipped with a g-suit to prevent the loss of blood in the brain, can withstand up to 9 g.[38] Autonomous drones will have no such constraints. In addition, when AI replicates human intelligence, autonomous weapon systems will be able to perform complex missions and replace humans in harm's way. For example, US Air Force drones will no longer require remote piloting. Instead, they will receive mission objectives, similar to human pilots, and on their own determine the best way to accomplish those objectives.

Let us summarize the weapons reality.

- The United States is currently operating under DoD Directive 3000.09, which prohibits the deployment of autonomous weapons unless commanders and operators are able to "exercise appropriate levels of human judgment over the use of force." With this constraint, the weapon can be termed semiautonomous.
- This directive does not prohibit the development or deployment of humans on-the-loop autonomous weapons.
- Our most advanced adversaries, China and Russia, have not imposed a directive similar to 3000.09 on themselves and are engaged in the development and deployment of autonomous weapons.
- All semiautonomous and autonomous weapons employ smart agents.
- We should expect the emergence of autonomous weapons with human-level autonomy in roughly the 2040–2050 timeframe.

Unless the United Nations is effective in its efforts to get nations to agree to ban autonomous weapons, the complexity of battle in the coming decades, combined with the fog of war, will necessitate that if any major player, like Russia, widely deploys autonomous weapons, China and the United States will also need to deploy them to maintain military parity.

THE COMPLETE PICTURE

I have attempted to assemble a collage of the new reality by looking at its component pieces, namely, 1) the Political Reality, 2) the Technological Reality, and 3) the Weapons Reality. In doing so, I have used modifiers such as "likely," "about," "may," etc. I now intend to provide the complete picture that I think is going to represent the new reality. I will not use probabilities or attempt to mitigate assertions. I will simplify my view by presenting it using only four bullet points followed by a short observation regarding the issues autonomous weapons will cause, especially as they relate to human annihilation:

- While the world is a tinderbox of tensions, and North Korea or another rogue state could ignite a limited nuclear war, the major powers (i.e., United States, China, and Russia) will not engage in an all-out nuclear confrontation, since such an exchange would likely result in human extinction.

- On or before the 2050, supercomputers will reach human-level intelligence, and superintelligence (the singularity) will emerge on or before 2080.
- The United Nations will fail to ban autonomous weapons.
- The United States, China, and Russia, among others, will deploy autonomous weapons with human-level autonomy on or before 2050.

In the new reality, autonomous weapons will present a danger to humanity beyond their destructive potential. First, their autonomy may ignite a war that humans would potentially avoid via diplomacy. When all major powers, and some rogue states, have autonomous weapons, the odds of a miscalculation or misinterpretation increases. Therefore, the odds of war occurring increases and that war could proceed on autopilot leading ultimately to human annihilation. Second, when autonomous weapons displace humans in harm's way, war may become more palatable. As the thought of war becomes more palatable, the probability of war increases. As the probability of war increases, the potential for human annihilation also increases. Lastly, at the point of the singularity, superintelligence may view humanity as a threat, given our history of war and the release of malicious computer viruses. Armed with autonomous weapons, superintelligence may wage war on humanity, causing us to fall victim to our own invention.

PART II

THE SECOND GENERATION: GENIUS WEAPONS

Our technological powers increase, but the side effects and potential hazards also escalate.

—Alvin Toffler, *Future Shock*, 1970

CHAPTER 5

DEVELOPING GENIUS WEAPONS

Previously, we discussed the emergence of autonomous weapons, with human-level intelligence, on or before 2050. I am not suggesting all weapons will incorporate human-level autonomy by 2050. I am asserting that these types of autonomous weapons will be one of many arrows in a technologically advanced nation's quiver. This is the norm. Advanced weapons emerge and become part of the overall arsenal of a nation, alongside current less advanced weapons. For example, the US Navy plans to deploy the USS *Gerald R. Ford* supercarrier in 2020. It will be the most advanced supercarrier deployed by any nation. However, the US Navy will still deploy less advanced supercarriers like the USS *Carl Vinson* until they reach the end of their service life. My point is that in 2050 expect to see a mix of weapons in the arsenals of the most advanced military powers. Also, expect to see a trend toward autonomous weapons.

Based on the Law of Decreasing Cost Returns (i.e., as technology increases, the cost of former generations of that technology decrease), expect that almost all weapons eventually will embody AI technology, from a gun that will only fire when held by its owner to autonomous fighter aircraft that perform complex combat missions. Also, along with the widespread introduction of AI technology in weapons, expect the level of interconnectivity to increase. This trend in the military will mirror the commercial "Internet of Things" trend (i.e., connecting any device to the

internet and to each other, from thermostats to washing machines) discussed in chapter 2. In the US military, widespread AI interconnectivity
will enable "swarming," a tactic that the military will borrow from nature
to bring all necessary resources together to ensure the defeat of an adversary. Germany used this tactic in World War II. The term "wolfpack"
was a German naval tactic that referred to the mass-attack tactics against
convoys used by German U-boats during the Battle of the Atlantic. The
United States used a similar tactic against Japanese shipping in the Pacific
Ocean. Although neither Germany nor the United States referred to this
tactic as swarming, in a more general sense it was. As the interconnectivity of weapons increases, expect the use of swarming to emerge as a
major military tactic. If this appears farfetched, let's consider an existing
example, namely, US Navy swarmboats. In 2014, the US Navy published a
media release, "The Future Is Now: Navy's Autonomous Swarmboats Can
Overwhelm Adversaries." In it, they assert:

> As autonomy and unmanned systems grow in importance for naval oper-
> ations, officials at the Office of Naval Research (ONR) announced today
> a technological breakthrough that will allow any unmanned surface
> vehicle (USV) to not only protect Navy ships, but also, for the first time,
> autonomously "swarm" offensively on hostile vessels.[1]

An example of one of the US Navy swarmboats in provided in figure 4.
Even with the gun clearly visible, the swarmboats don't appear particularly destructive compare to a larger warship like a destroyer. However,
appearances can be deceptive. If you see one ant or one bee, you will likely
have little concern. On a picnic, you may even swat it if it becomes annoying.
Now envision a swarm of army ants or killer bees invading your picnic. Likely
your first instinct, rather than swatting at them, would be to get yourself and
your loved ones to safety. What makes them dangerous is their swarming.
The US Navy's swarmboats apply the same principle. Their intent is to
guard larger warships while in port from an adversary's smaller boats that
could potentially damage or even sink the larger warship. This requirement
becomes clear if you consider the incident of the USS *Cole*.

The USS *Cole* is a US Navy guided-missile destroyer. Obviously, the
Cole is a formidable warship. However, on October 12, 2000, the USS *Cole*
docked in Aden harbor, in Yemen, for a routine fuel stop. During the refu-

eling, a small fiberglass boat carrying explosives and two suicide bombers approached the port side of the *Cole*. The small craft exploded as soon as it reached the destroyer, creating a forty-by-sixty-foot gash in the ship's port side. The attack killed seventeen American sailors and injured thirty-nine others.[2] The US Navy is developing swarmboats to counter similar threats.

Figure 4. Newport News, VA (Aug. 12, 2014): An unmanned eleven-meter rigid-hull inflatable boat from Naval Surface Warfare Center Carderock operates autonomously during an Office of Naval Research–sponsored demonstration of swarmboat technology on the James River in Newport News, Virginia. During the demonstration as many as thirteen Navy boats, using an Office of Naval Research–sponsored system called CARACaS (Control Architecture for Robotic Agent Command Sensing), operated autonomously or by remote control during escort, intercept, and engage scenarios. (US Navy photo by John F. Williams/Released)

Initially, in 2014, a human operator had to tell the robotic swarm-boats which vessel to swarm. However, by 2016, with advances in AI, the swarmboats could autonomously determine friend from foe. Its recognition technology can even evaluate potential threats based on behaviors. The technology will pay attention to how close a suspect vessel is getting to a Navy asset. In addition, the tactical capabilities of swarmboats have also improved. Instead of all swarmboats swarming one vessel, which could leave a potential opening for another adversary's vessel to attack, the US Navy's swarmboats can assign different drones to track different enemy boats.[3]

This example illustrates three points:

1. Increases in AI technology can be a game changer.
2. Distribution of AI technology can enable swarming attacks.
3. Miniaturization, such as nanoelectronics, can enable new weapon capabilities and potentially a new class of weapons.

Let me elaborate on the last point, "miniaturization." In chapter 3, we discussed that US military's Third Offset Strategy. The Third Offset Strategy targets several promising technology areas to develop, which would allow the deployment of weapons exploiting robotics and system autonomy, miniaturization, big data, and advanced manufacturing. Obviously, the US Navy's development of swarmboats is an excellent example of the Third Offset Strategy. Swarmboats are robotic drones capable of autonomy. They are small compared to the US Navy's typical warships. They incorporate advanced AI and access big data to autonomously identify friend from foe. Due to their small size and limited capabilities, advanced manufacturing could produce them quickly and cost effectively. Imagine a fleet of swarmboats prowling the coastal waters of any nation. They could conceivably keep that nation from deploying its largest warships. Imagine swarmboats equipped with anti-ship missiles. Their speed and large numbers would make them the Navy's killer bees.

While the United States will continue to deploy large weapon systems, like supercarriers, I also expect it will focus on smaller weapon systems, like swarmboats. For example, I expect that, in addition to large fighter drones, the US military will deploy insect size drones. The drive to miniaturize weapons and endow them with advanced AI capabilities will eventually give rise to an entire new class of weapons, namely, nanoweapons.

In chapter 3, I mentioned that we could interpret miniaturization to include nanoweapons. Nanoweapons, as defined in chapter 3, are any military technology that exploits nanotechnology (i.e., technology with at least one-dimensional element between one and hundred nanometers). While I explicitly asserted that cyberwarfare likely relied on nanoelectronic microprocessors, making it an example of nanoweapons, I did not point out other potential applications. This was intentional because I felt it would be better to discuss them here to set the foundation to discuss genius weapons.

The first weapon system we discussed in chapter 3 was the Aegis Combat System. The US Navy has deployed this system since 1983. In its current configuration, it likely includes computer technology dating back three to five years or more, as well as the most advanced computer technology. In those applications where the Aegis Combat System is using the most advanced computers, they include nanoelectronics microprocessors. Therefore, we should classify those portions of the Aegis combat system as belonging to nanoweapons category 3—Defensive Tactical Nanoweapons. At first, it may seem odd to make this assertion. I agree. In fact, looking at Aegis as a weapon is odd. Isn't it just computers, algorithms, radar, and the like? Nothing in it actually goes "bang." However, it directs US missiles against threats and those missiles do go bang. Typically, the bang is large enough to take out a naval vessel. My point is that as we discuss genius weapons, a paradigm shift will be necessary. Many of them will not look like today's conventional weapons. Even when they do, the nanoweapon aspects are subtle. For example, let examine the second weapon we discussed in chapter 3, the X-47B Unmanned Combat Air System.

The X-47B Unmanned Combat Air System fits the picture we typically conjure in our mind's eye when we envision an advanced weapon system. Is this a nanoweapon? The US Navy is not making public the technology used to build the X-47B. However, given its high-tech semiautonomous capabilities, I judge that it uses the latest nanoelectronic microprocessors. This allows us to classify it as a nanoweapon, and it belongs to category 2—Offensive Tactical Nanoweapons—as well as category 3—Defensive Tactical Nanoweapons. But, there may be more. It's also possible it uses nano-based materials in its manufacture and as part of its stealth coatings, but the secrecy surrounding its technology makes this difficult to ascertain.

My point is that nanotechnology is an enabling technology. When it enables a military system, that system becomes a nanoweapon. I recognize this is a broad definition, but it is also necessary and accurate. If you doubt this, try removing the nano-based products from the system. This would mean that the Aegis Combat System could only use computer technology dating back before 2011, the date when nanoelectronics became available. A computer in the 2011 vintage would exhibit processing power ten times less than a thousand-dollar midrange computer that you could procure at any consumer electronics store. Do you think the US Navy would allow Aegis, one of its most important combat systems, to become that antiquated? Let's take another example. Try removing the nano-based products from the X-47B. Without its nanoelectronic processors, I doubt it would be semiautonomous. Although the US military doesn't flaunt it, nanotechnology is embedded in its most advanced military systems. By definition, they are nanoweapons.

In my book, *Nanoweapons: A Growing Threat To Humanity*, I made two important points:

1. The US military decisively leads its most technologically advanced adversaries in nanoweapons.[4]
2. The US military is intentionally keeping this quiet.[5]

You may wonder why the US military is being secretive regarding its use of nanotechnology and why you don't hear more about nanoweapons. The answer is deceptively simple. The United States decisively leads in nanotechnology and its application to weapons, which gives it a strategic advantage. It is intentionally downplaying any mention of nanoweapons. My book *Nanoweapons* was the first to bring the existence of this new class of weapons to the lay public's attention.

The following are the main elements discussed in *Nanoweapons* that pertain to genius weapons, along with some of their implications. After establishing this foundation, we will be in a position to define genius weapons.

Nanoelectronic Integrated Circuits

As we discussed previously, artificial intelligence technology is at the core of the US military's Third Offset Strategy. The lifeblood of artificial intelligence technology is integrated circuits and, in its more advanced aspects, nanoelectronics.[6] Smart weapons require artificial intelligence technology, and it plays numerous roles. AI provides the guidance system of smart weapons. AI can also trigger when a smart weapon detonates. For example, AI technology may delay detonation until a weapon penetrates a barrier to ensure it neutralizes all enemy combatants behind the barrier or destroys a weapon within a bunker. As advances in nanoelectronics result in advances in AI technology, expect almost all weapons gradually to become smarter. Also, expect nanoelectronics to enable smart versions of small munitions, such as self-steering smart bullets that enable snipers to make one-mile kills or to target a specific individual based on their DNA.[7] Most importantly, as discussed in chapter 4, expect nanoelectronic integrated circuits to enable human-level intelligence on or before the year 2050. In addition, as discussed in chapter 4, superintelligence will occur on or before 2080. While I recognize my projections may be off by a decade or longer, superintelligence is still likely to emerge in the latter portion of the twenty-first century.[8] When this occurs, the world will experience a singularity. We will have a superintelligence whose performance greatly exceeds the cognitive performance of humans in virtually all domains of interest. As we incorporate superintelligence in weapons technology, we will witness the emergence of genius weapons.

Nanosensors

Sensors are a critical technology employed in smart weapons. However, nanosensors are an entire new class of sensors and will be critical to developing and deploying genius weapons. To provide a deeper understanding, let me quote from my book *Nanoweapons*: "Nanosensors offer unparalleled opportunities to interact (i.e., sense) at the molecular level. This makes them extremely effective as biosensors and chemical sensors, where the requirement is to detect low concentrations with high specificity."[9]

Nanosensors enable AI control systems to be highly capable in combat scenarios. For example, the self-steering smart bullets we discussed in the previous section requires nanosensors. Using nanosensors, conceivably these bullets could be capable of facial recognition. This may change the old adage "a bullet with your name on it" to "a bullet with your picture in it." Since nanosensors interact at the molecular level, they will enable AI control systems to distinguish between combatants—people carrying weapons and explosives—from noncombatants. They will do this via "the detection of explosives, biological warfare agents, and chemicals."[10] Imagine a US soldier in a combat zone confronted by a number of locals, including women and children. One or more may carry weapons. However, the soldier has no way to knowing which, if any, are enemy combatants. In this scenario, imagine the soldier firing smart bullets with nanosensors capable of detecting those with weapons. If one or more locals have weapons, the bullets will strike only them. If no one is carrying weapons, the bullets will self-steer and miss all locals. I still classify this as a smart weapon, but we are moving closer to genius weapons. As will become apparent later, when we define genius weapons, some will require nanosensors.

Nanorobotics

Robotics is playing an increasing role in warfare, from surveillance to offensive operations.[11] Previously, we talked at length about US military drones, such as the US Air Force's remotely piloted MQ-1 Predator drone and the US Navy's semiautonomous swarmboats. We also discussed how drones will become more autonomous as artificial intelligence emulates human intelligence, which in chapter 4 I projected to occur on or before 2050. While this will represent a significant advancement in military robotics, I still would not classify them as genius weapons. Nanorobotics will further increase the role of military drones. Imagine a fly drone capable of surveillance within an adversary's command center. This would give a completely new meaning to the phase "fly on the wall," Imagine the fly drone depositing 100 nanograms of botulism H, the most deadly toxin in existence and with no known antidote, on the food of a high-value enemy combatant, such as a theater commander. This may read like science fiction, but on December 16, 2014, the Army Research Laboratory announced the creation of a fly drone,[12] and botulism H is also real as described.[13] Because of the unparalleled capabilities that nanorobotics offers, when we define a genius weapon it will become evident that nanorobots will play a key role. In general, robotics will play a central role in genius weapons.

These three elements—nanoelectronics, nanosensors, and nanorobotics—provide a foundation for us to define a genius weapon. The phrase "smart weapon" connotes an artificially intelligent, precision-guided weapon, which has extraordinary accuracy, minimizes collateral damage, and increases lethality against intended targets. We began to use the phrase during the Gulf War. In time, the word "smart" became a synonym for AI. Thus, when our cell phone embedded AI, we called it a smartphone. Today, we use the phrase "smart weapon" to mean a weapon that embeds AI and emulates tasks normally requiring human intelligence. For example, the X-47B Unmanned Combat Air System is able to take off and land on an aircraft carrier. In general, as weapons approach capabilities we normally ascribe to humans, we will label them "smart." However, when a weapon's performance greatly exceeds the cognitive performance of humans in virtually all domains of interest, I judge that we will call it a "genius" weapon. With this background, let us define the attributes that constitute a genius weapon. This will not be an official military definition, no more than "smart," as it refers to weapons, is an official military definition. This definition will define three weapon attributes that will cause us to term them genius weapons:

1. It must be a robotic weapon that either embeds superintelligence or is able to wirelessly connect to superintelligence.
2. It must be controllable and able to carry out designated military missions. Based on the first attribute, this means it will be under the control of superintelligence, which presumably is under the control of human commanders.
3. It must be able to cause various levels of destruction, from destroying an adversary's largest weapons, like supercarriers and nuclear-tipped missiles, to annihilating an army or killing a single enemy combatant.

This definition may appear arbitrary and seem to be impossible for any weapon to meet all criteria. It is not arbitrary. Two categories of genius weapons fit these criteria.

I. ROBOTIC WEAPONS THAT ARE WIRELESSLY CONTROLLED BY SUPERINTELLIGENCE

In the fourth quarter of the twenty-first century, two critical capabilities will intersect to provide a genius-level weapon. Those capabilities are:

1. The existence of superintelligences
2. Significant advances in nanorobotics, including nanoelectronics and nanosensors

Clearly, the genius weapon that fits the above definition is militarized autonomous nanobots (MANS).

Nanobots refer to tiny robots, made using nanotechnology. Typically, in the latter portion of the twenty-first century, nanobots will incorporate nanoelectronics and nanosensors that give it AI functionality and the ability to wirelessly communicate with superintelligences. With these capabilities, nanobots become autonomous and move from "smart" to "genius." Such nanobots would be capable of performing military missions. Thus, we can term them militarized autonomous nanobots (MANS). You may wonder, though: Why would MANS constitute genius-level weapons? Actually, this is a logical extension of how the United States and other nations conduct war. Currently, the United States makes extensive use its Global Positioning Satellites to guide drones and smart bombs during conflict. Military robotics is now an indispensable technology that the United States and other countries use to make war.[14] Asserting that MANS will be a genius weapons results from extrapolating the current military trend regarding shrinking drones to the size of insects,[15] along with the role robotics plays in warfare. This may become clearer if we consider a potential mission for MANS. Imagine, for example, a swarm of millions of MANS, with resident nanoelectronic and nanosensors, being directed by superintelligence to deliver a deadly toxin to a nation's adversaries. In this example, MANS are mimicking a biological plague. The point of this example is to demonstrate that MANS, like nuclear weapons, are strategic weapons of mass destruction.

In order for MANS to be genius weapons, capable of performing a wide variety of tasks, such as conducting a swarming attack, as well as responding and adapting to its environment, they must have superintelligence functionality. Given their size, it most likely will not be possible to build that level of AI functionality into MANS. Their genius attribute would rely on the superintelligence that wirelessly controls them. As nature gave bees the instinct to swarm to attack enemies, superintelligence would wirelessly coordinate the functions of the millions or billions of MANS engaged in a swarm attack.

Lastly, for MANS to be able to destroy an adversary's largest weapons or it combatants, they must exist in large numbers. Based on the mission, millions to billions of MANS may be required. For example, if we wanted the MANS to attack an aircraft carrier, each militarized autonomous nanobot could carry a minute corrosive payload. When it reached the air-craft carrier, it would release its corrosive payload. As millions to billions of MANS attack the carrier, conceivably they could cause the carrier hull to lose its integrity, which would cause the carrier to sink. Initially, MANS may be manufactured and transported to the theater conflict. While delivery of large quantities of MANS into the theater of conflict is possible, it adds a level of difficulty and vulnerability. Taking a page out of nature's playbook, eventually we would want MANS to be self-replicating. Self-replication removes the vulnerability. Consider this: The bubonic plague would not have been able to kill tens of millions if did not have the ability to replicate. In modern times, quarantining people with an infectious disease is one method to keep an infectious disease from replicating, which we normally term "spreading."

In *Nanoweapons*, I categorized self-replicating MANS as strategic offen-sive nanoweapons of mass destruction.[16] These weapons are more capable in their selective destructive capability than nuclear weapons. Self-repli-cating MANS will be able to carry out extensive military operations, from surgical strikes that kill a single enemy combatant to massive attacks capable of destroying a nation's largest weapons, such as aircraft carriers and sub-marines. The destructive potential of self-replicating MANS requires that their use always be controlled. Given the complexity of controlling mil-lions to billions of MANS, superintelligence would be necessary to control them. However, this raises a significant question: Will humanity be able to maintain control of superintelligences? We will address this question in the next chapter.

The whole concept of nanobots may come across as science fiction, but they are not. Nanobots exist today. While the military classifies its man-ufacture and use of nanobots, the medical profession openly discusses it. To give credence to this whole area, let's digress and look at some of the medical applications of nanobots.

On January 6, 2016, the Leukemia Research Foundation announced, "Ido Bachelet, PhD from Bar Ilan University in Israel, is conducting the first human clinical trial using nanobots to treat late stage leukemia."[17] They referenced two articles:

1. Brian Wang, "Pfizer Partnering with Ido Bachelet on DNA Nanorobots," *Next Big Future* (blog), May 15, 2015.

   According to the article, "Pfizer is cooperating with the DNA robot laboratory managed by Prof. Ido Bachelet at Bar-Ilan University. Bachelet has developed a method of producing innovative DNA molecules with characteristics that can be used to 'program' them to reach specific locations in the body and carry out pre-programmed operations there in response to stimulation from the body."[18] In summary, Dr. Bachelet team makes a nanobot by using a specific DNA sequence and folding it like a "clam" so that it can hold cancer-fighting drugs. This specific-DNA clam-like molecule travels inside the patient's body until it finds a cancer cell. At that point, it releases the drug to kill the cancer cell. In this way, the cancer fighting drugs only target cancer cells. Unlike conventional cancer treatments that can attack healthy cells as well as cancerous cells, this method only attacks cancer cells.

2. Daniel Korn, "DNA Nanobots Will Target Cancer Cells in the First Human Trial Using a Terminally Ill Patient," *Plaid Zebra*, March 27, 2015.

   This article provides additional information, stating, "Nanobots can also have multiple 'payloads' in them, and can be programmed so that they know which drug to expose to specific molecules."[19]

The medical nanobots discussed in the above articles use DNA molecules. While these are biological nanobots, medical nanobots' research is taking place on a worldwide basis, and some research includes technological nanobots. For example, China's Harbin Institute of Technology is using nanobots with magnetic rotating arms.[20] These nanobots use their rotating arms to swim in a patient's blood vessels. Researchers apply a magnetic field to activate the nanobots' arms. The nanobots then reach a specific destination to administer medicine from inside the patient's veins.

This short segue into medical nanobots makes an important point. Nanobots exist today. This is science fact, not science fiction. There are more examples. A Google search on August 16, 2017, for the keyword phrase "medical nanobots" (with the quotes) yielded 83,200 search returns. These include articles from the *Atlantic* and *Business Insider*. By contrast, a Google search on August 16, 2017, for the keyword phrase "military nanobots" (with the quotes) yielded 2,410 search returns (around 3 percent of that for "medical nanobots"), and none of the results appear to come from the US military. However, the US military is spending billions on nanoweapons.[21] This leads me to conclude that they are at least exploring the use of nanobots for military applications. However, having worked on classified programs, I know those working on military nanobots cannot publish their research or report it in a public conference, which explains why our Google search for "military nanobots" yields so few search returns.

Before we leave the subject of nanobots, let me mention that they have one capability both modern society and the military will use. That capability will be molecular manufacturing. K. Eric Drexler, one of the founders of nanotechnology, described it this way:

> Molecular manufacturing—although in most respects quite different from anything biological—will likewise use stored data to guide construction by molecular machines, greatly extending abilities in nanotechnology.[22]

Drexler observed that nature built large biological machines using cells at the nano level.[23] It turns out that all biological processes begin at the nano level. Mother Nature's tool of creation is nanotechnology. The DNA provides a roadmap that enables biological cells to form larger structures, such as hearts and kidneys, and eventually a complete animal, which could be a human or a dog.

Physicist and Nobel Laureate Richard Feynman gave a lecture at the American Physical Society meeting at the California Institute of Technology on December 29, 1959, titled, "There's Plenty of Room at the Bottom."[24] In his lecture, Feynman introduced the concept of molecular manufacturing. Although Feynman never used the words "nanotechnology" or "molecular manufacturing," he described a process involving the precise manipulation and control of individual atoms and molecules to build nanomachines. He further envisioned factories using nanomachines to cost-effectively build complex products, even manufacturing products impossible using today's conventional manufacturing methods.

Just as billions of termites can eat away at a mansion made of wood, leaving no trace, nanobots in reverse can build a mansion or any structure. That structure could be a complex machine or a building. One important point is that the resulting structure will be precise. According to plan, each atom and molecule will be precisely in place, as opposed to today's manufacturing methods, which at the atomic level are grossly irregular.

Two questions naturally present themselves:

1. Is this possible?
2. Is it important?

As I mentioned, everything Mother Nature builds starts at the nanoscale. Therefore, we know such building is possible, from the smallest viruses to the largest coral reefs. However, you may correctly point out that this has to do with biology. What about non-biological structures? Are they possible? The answer is yes, although they are only now becoming feasible.

Nicholas Kotov and his colleagues at the University of Michigan recently published a technique to build larger nanostructures using a process that builds materials one nanoscale layer at a time, similar to the way Mother Nature builds abalone shells in layers.[25] Kotov's process consists of dipping a piece of glass about the size of a stick of gum into a glue-like polymer solution and a dispersion of clay nanosheets (i.e., a two-dimensional nanostructure of layered clay minerals), which form hydrogen bonds with each other across the layers. Kotov's team repeats the process to make a strong final product about as thick as a piece of plastic wrap. Although this is still in the early phases of development, it is conceivable that eventually larger structures made with this process could become the new "plywood" in construction.

Why is this important? We find that exact atom or molecule placement can result in extremely strong structures. Although an abalone shell consists of 98 percent calcium carbonate, thanks to its precise arrangement of molecules it is three thousand times stronger than calcium carbonate rocks.[26] Nanotechnology is already affecting the largest structures in our world. For example, we know that when we add carbon nanotubes (i.e., molecules of pure carbon that are long, thin, and shaped like tubes) to concrete, it becomes stronger. This occurs because the carbon nanotubes influence the processes of hydration and the subsequent crystallization of new products in the cement paste.[27] In *Nanoweapons*, I describe how specifically constructed nanosheets are strong enough to stop bullets.[28] The point is that nanomanufacturing is already beginning to revolutionize manufacturing, enabling the creation of products that otherwise would not exist.

The bottom line is that nanobots are not only useful as a candidate for genius weapons, but they could profoundly change medical treatments and manufacturing. This is a subject worthy of its own book, and given the scope of this work I can only touch on it to provide an appreciation for the potential.

II. ROBOTIC WEAPONS THAT HAVE SUPERINTELLIGENCE EMBEDDED AS PART OF THE WEAPON

Given how the military is operating today, I judge that the second category of genius weapons will include a nation's largest weapons, aircraft carriers and submarines. Imagine such a weapon system being completely automated and housing superintelligence. This weapon system completely fits our definition of a genius weapon. It's robotic, controlled by superintelligence, and capable of various levels of destruction. In addition to launching missiles and drones, it will also have the capability to launch MANS.

One characteristic the United States excels in is projecting force. Today, two ways we project force is via our nuclear aircraft carriers and submarines. In the future, we will continue to project force via nuclear aircraft carriers and submarines, but the nuclear aircraft carriers and submarines will be robotic. This carries a trend we are seeing today to its conclusion. For example, due to the increased level of automation, our most advanced Ford-class carrier has a much smaller crew than the Nimitz-class carrier. With the development of superintelligences, all operations aboard a carrier or submarine are automatable. Imagine a nuclear submarine that could stay on missions measured in years not months.

In addition, supercarriers and nuclear submarines would be able to bring MANS into a theater of conflict, as well as conventional and nuclear weapons.

As superintelligences emerge in the latter part of the twenty-first century, they will design weapons that today are the subject of science fiction. However, they will not only design weapons that no longer require human operators, they will replace humans in almost every aspect of human endeavor.

Fueled by an "intelligence explosion," where each generation of intelligent machines develops the next and even more intelligent generation of machines, nations with superintelligences will experience an exceptionally high standard of living, as these machines alleviate all human suffering and manufacture an abundance of food and products fulfilling every human need. These nations will also have an unprecedented level of military capability due to the genius weapons in their arsenals.

Autonomy of supercarriers and submarines will raise another question: What role will humans play? In the latter part of the twenty-first century, autonomy of supercarriers and submarines will be in its early phases. I believe that humans will choose an on-the-loop role. While these robotic machines will be able to operate with full autonomy, human commanders will likely want to have knowledge of their activities, especially as related to combat, and have the ability to override their autonomous decisions.

Obviously, any conventional communications may provide intelligence to adversaries regarding the position of such genius weapons. Such communications would also vulnerable to hacking. Imagine a terrorist hacker gaining control of a genius-level submarine. Given its capabilities, that scenario could spell human annihilation. Therefore, it will become imperative that the US military, and other militaries, be able to communicate with their genius weapons without allowing adversaries to learn their positions or intentions or to gain control over them. What kind of communication technology can fill this need? The answer, I judge, will be at the heart of how superintelligence works.

As we discussed in chapter 1, Moore's law demonstrates that integrated circuits cost-effectively double in density every two years. However, we are currently making integrated circuits with features in the nanoscale range. This suggests to me that the traditional methods employed to make integrated circuits will reach a point where improvements in device dimensions on integrated circuits will hit a wall. Yet we also asserted in chapter 1 that Moore's law is an observation of human creativity in any well-funded field of technology, such as computer technology. Therefore, even as conventional integrated circuits reach their limit of performance improvements, computers may continue to improve exponentially, fueled with new technology. Just as transistors replaced tubes and integrated circuits replaced transistors, we are likely to see quantum computing emerge to replace integrated circuits. Using this new technology, improvements in computer capabilities will continue to follow Moore's law, or as we more generally stated it, the Law of Accelerating Returns.

Naturally, this raises a question: What is quantum computing? Quantum computing refers to computers that use quantum-mechanical phenomena, such as entanglement, to perform operations on data. Einstein termed quantum entanglement, "spooky action at a distance."[29] Entanglement occurs when two particles interact and their properties become interdependent. For example, let's assume that two subatomic particles interact, resulting in the emergence of two electrons, created at the same point and instant in space. In the language of quantum mechanics, the electrons are "entangled." This means that a measurement on one electron immediately influences the other, regardless of the distance between them. If one of the electrons is in a state of "spin up," a characteristic of its angular momentum, the other will be in a state of "spin down," to conserve spin (i.e., a principle of quantum mechanics). If you change the spin of one electron from spin up to spin down, the other will immediately change from spin down to spin up to conserve spin. Quantum entanglement can occur between groups of subatomic particles, depending on their creation and interaction. Although this may seem impossible, quantum entanglement is widely accepted by the scientific community as a fundamental feature of quantum mechanics. From the standpoint of a computer, whose language is ones and zeros, the electron's state—for example, spin up—could represent a 1, and spin down could represent a 0. This may come across as science fiction, but it is not. It is science fact.

In August 2016, China launched a satellite, known as Quantum Experiments at Space Scale, from the Jiuquan Satellite Launch Center.[30] China's new satellite is able to emit quantum particles such as photons to send secure information. This is termed quantum cryptography. This works by altering the characteristics of the photons to generate a key of 1s and 0s and then transmitting the key (the photons) to the intended recipient. The recipient is then sent an encrypted message, again via photons, which they can decipher with the key. While this is in the early stages of development, it is proof of concept.

Quantum cryptography uses another principle of quantum mechanics that makes it theoretically impossible to hack. That principle is the Heisenberg Uncertainty Principle, which states that when you try to measure the state of a subatomic particle you disturb its state. This means that if a potential hacker tries to steal the key in transit, they would instantly change it to a different set of states (i.e., different 1s and 0s). In theory, this makes quantum cryptography impossible to hack. How do the Chinese intercept the key and use it to decode the message, since in theory they would also disturb the quantum states? The Chinese are not telling, but I suspect they are using error-correcting code (ECC) resident on their intercept computer. ECC is a type of code that can detect and correct internal data corruption. I recognize that this may be getting highly technical. It suffices to say that the Chinese have figured out how to use a form of quantum coding, which is a building block of quantum computing.

Using the properties of atoms and subatomic particles will likely represent the next leap in computing technology. Therefore, it's reasonable to conclude that superintelligences will be quantum computers. The internal operation of a quantum computer will utilize the principles of quantum mechanics, which is the science that seeks to explain the behavior of atoms and subatomic particles at the quantum level. Currently, the science of quantum computing is in its infancy, with development occurring in highly advanced laboratories. Fortunately, for our purposes, we really only need to understand the three major advantages of quantum computing:

1. Quantum computers are inherently faster than classical computers (i.e., our current electronic computers). For example, a classical computer uses electrons and photons to perform calculations. Although those electronics and photons may traverse a computer's internal circuits in a small fraction of a second to transfer information, they are not instantaneous. By comparison, quantum computers use the properties of subatomic particles, such as quantum entanglement. Some researchers argue the transfer of information between entangled particles appears to occur instantly, indepen-

dent of distance.[31] If true, quantum computers would be significantly faster than today's computers, which use conventional interconnected circuits.

2. Quantum computers can tackle more difficult problems than classical computers. This is a direct result of using the properties of subatomic particles, which have the strange ability to exist in more than one state at any time. You can think of this as adding an additional level of information, allowing quantum computers to make complex calculations that are beyond the practical scope of classical computers.

3. Quantum computers use less energy than classical computers. A classical computer requires a theoretical minimum amount of energy to perform one operation, which IBM Research Lab's Rolf Landauer calculated in 1961.[32] Our current computers use millions of times more energy than Landauer's minimum. In this respect, they are not energy efficient. Researchers believe quantum computers will be more energy efficient than classical computers, even if classical computers approach Landauer's minimum. For example, changing an enormous amount of information in a quantum computer may only require changing the state of one subatomic particle. When one subatomic particle's state changes, it can influence the state of numerous other particles via quantum entanglement.

Quantum computers will usher in a new level of computer processing power, and they will likely be the machines that usher in the singularity, the point in time when intelligent machines greatly exceed the cognitive performance of humans in virtually all domains of interest. By definition, genius weapons will utilize superintelligence to attain their genius-level capabilities. As they emerge in the latter part of the twenty-first century, we will be dealing with intelligences greater than human intelligences. Controlling genius weapons implies that human will have control over superintelligences. This raises a serious questions: Will superintelligences accede to human control? Alternatively, will they take a different view of their place in the food chain?

If we continue to develop our technology without wisdom or prudence, our servant may prove to be our executioner.
— Omar N. Bradley, Armistice Day speech, 1948

CHAPTER 6
CONTROLLING AUTONOMOUS WEAPONS

This chapter discusses the difficulties humans are likely to encounter in controlling autonomous weapons as both humans and AI technology advance.

CONTROLLING AUTONOMOUS WEAPONS WITH WEAK AI

In general, you can think of "weak AI" as artificial intelligence up to, but short of, human-level intelligence. It is the type of AI we currently utilize to play sophisticated computer games. The word "weak" simply means it is not human-level intelligence. When we talk about "strong AI," we will be talking about AI that is equivalent to human-level intelligence.

In the short term, and by short term I mean prior to the emergence of strong AI, autonomous weapons will remain controllable by their human operators. The United States will continue to abide by its self-imposed restriction to field only semiautonomous weapons until adversaries field autonomous weapons that place US military forces at a disadvantage. Russia and China are the most likely adversaries to deploy advanced autonomous weapons, which will give them an advantage over humans in combat. As human casualties increase in conflicts against autonomous weapons, there will be a strong reaction within the US populace to deploy autonomous weapons, essentially a "fight fire with fire" approach. Why lose sons and daughters, husbands and wives, friends, and others to autonomous weapons?

I expect semiautonomous weapons to be highly effective until autonomous weapons begin to achieve human-level intelligence. At that point, autonomous weapons will replace humans on the battlefield and will be more capable of adapting to changing circumstances. As autonomous weapons with human-level intelligence emerge, the United States will modify its stance on autonomous weapons. The United States may claim their weapons are semiautonomous, with a human on-the-loop control. However, the pace of conflict will relegate the human commanders to essentially the role high-level commanders played in World War II.

During World War II, the theater commander was responsible for conducting the war in the theater of conflict (for example, the European theater). He ordered the destruction of specific targets, deemed necessary to deprive the adversary of a critical capability, from weapon factories to bridges. Planning how to achieve the theater commander's orders became the responsibility of his staff. After the theater commander approved their plan, high-level commanders in each of the services would delegate various elements of the plan to their staff. They, in turn, would plan specifically how to achieve their assignments. Once all plans were in place, a synchronized attack would commence. Each officer would lead the men under his command to accomplish their assigned tasks. As the attack unfolded, officers in the high command would continue to monitor and report progress. The theater commander could modify the plan, if time and communications permitted, to address the adversary's counteractions. Often, though, such modification would come from the commanders on the ground, since they were closer to the events unfolding.

If you consider the above, the theater commander and his staff had three responsibilities:

1. Plan and communicate.
2. Monitor and report.
3. Modify the plan based on the adversary's actions.

I recognize the above is a simplification. Obviously, there were many interactions at all levels during the planning and its implementation. However, from a broad-brush perspective, it is accurate. In today's parlance, we could argue that high-level commanders were on-the-loop, not in-the-loop. The high commanders did not pull triggers or drop bombs. Often that task fell to those in the lowest levels of the command chain, who had no knowledge of the larger plan. In effect, they were following orders.

When we use autonomous weapons, with humans on-the-loop, is this any different from the World War II scenario above? For example, the theater commander will develop a plan, interacting with his staff. The theater commander will approve the plan. His staff will meet with their staffs, similar to the World War II scenario above. The actual release of autonomous weapons will be the responsibility of troops under their command. The responsibility to implement the plan will fall to autonomous weapons. Once again, commanders at all levels will be on-the-loop, assuming the same type of responsibilities their predecessors assumed in World War II. However, the pace of conflict may increase to the level at which no modification or intercession on the part of the human commanders may be possible, short of aborting the mission. The autonomous weapons will make modifications to the plan as they encounter adversarial countermeasures. To that extent, the autonomous weapons will assume the same role as the commanders on the ground during World War II.

Viewed this way, the use of autonomous weapons may seem acceptable. They are simply implementing the war plans of humans. For example, when the US Navy fires its MK-50 torpedo at an adversary's fast, deep-diving submarine, its guidance system uses active and passive acoustic homing to make adjustments to counter any evasive tactics the adversary may employ. How is this different from autonomous weapons with humans on-the-loop?

In this type of scenario, there is no discernable difference. However, when autonomous weapons attain human-level intelligence (i.e., strong AI) the scenario may not be as depicted above. There are likely to be significant differences.

When AI technology becomes "strong," meaning it is equivalent to human-level intelligence, I mentioned there would be significant differences related to their control. To explore these differences, let us digress to consider two important questions:

1. Should we consider machines with human-level intelligence a new life-form, artificial life (i.e., Alife)?
2. If we consider human-level intelligent machines as Alife, what rights should they have?

These questions may seem odd, since we are dealing with machines and not biological life-forms. In the United States, and in other nations, we pass laws to protect lower life-forms, such as cruelty to animal laws, as well as laws to protect endangered species. In the case of human-level intelligence machines, we will have machines more intelligent than animals. At the point when a machine reaches the equivalent of human-level intelligence, is it no longer simply a machine? Should we enact laws protecting these machines, essentially establishing machine rights?

Proponents who hold that a machine with human-level intelligence is a life-form (i.e., Alife) on par with human life fall into the category termed "strong Alife," which stands for strong AI life (i.e., human-level intelligence life). A number of prominent people have been strong Alife proponents. For example, Hungarian-born American mathematician John von Neumann (1903–1957) held that "life is a process which can be abstracted away from any particular medium."[1] In the early 1990s, ecologist Thomas S. Ray made an unusual assertion. He claimed his Tierra project (a computer simulation of artificial life) did not simulate life in a computer but synthesized it.[2] Famous science fiction writer Arthur C. Clarke made this assertion in his novel *2010: Odyssey Two*: "Whether we are based on carbon or on silicon makes no fundamental difference; we should each be treated with appropriate respect."[3] These assertions all argue that life is independent of biology.

Given the above views, let us address question number one: Should we consider machines with human-level intelligence a new life-form, artificial life (i.e., Alife)?

Since these machines equate to human-level intelligence, it is reasonable to consider them a new life-form, Alife. Although they are an invention and made in factories, the final product is more than the sum of its parts. My answer is subjective, and I can respect those who hold a different view. However, let me be clear; I am not claiming Alife is equivalent to humans. Having human-level intelligence does not mean they are human. For example, they may not have the full spectrum of human capabilities, such as the ability to create or become emotionally involved. However, having human-level intelligence would suggest they know the difference between existing and not existing. Because of this, I argue they are not simply machines, they are life-forms. If you view this differently, however, I respect your viewpoint.

Let us now turn our attention to question two: If we consider human-level intelligent machines as Alife, what rights should they have?

In 2002, Italian engineer Gianmarco Veruggio coined the term "roboethics."[4] Roboethics focuses on the moral behavior of humans as they design, construct, use, and treat artificially intelligent beings. This raises a question: What are the moral obligations of society toward its artificially human-level intelligent machines? As previously mentioned, this question parallels the moral obligations of society toward animals. For intelligent machines with human-level intelligence, some may argue that machine rights should parallel human rights, giving them, for example, the right to life, liberty, freedom of thought and expression, and even equality before the law. However, if we take machine rights to that level, consider these consequences:

1. What legal means would we have to require intelligent machines serve us or fight wars as our proxies?
2. How would we control the evolution of intelligent machines?

The implication of the first consequence is that we would have to consider Alife our equals. We could not require them to serve or protect us. You may believe we could live with this consequence and that there would be harmony between humans and Alife. For example, humans now fight in defense of their country; perhaps Alife would adopt the same posture. Unfortunately, in the long term, I doubt that would occur. To understand this, let us explore the implications of consequence two.

If we consider Alife as having the same rights as humans, it would rule out controlling their evolution. Imagine enormous factories manufacturing millions of human-level machines. In a short while, they form a sizable population. Imagine those machines designing more advanced intelligence in machines as they manufacture each new generation of Alife. This is the definition of an intelligence explosion: Each generation of intelligent machines designing the next generation with more intelligence than the last generation. This means these machines would be equal to human-level intelligence in general and outperform humans in specific tasks. For example, they could have all the knowledge necessary to be neurosurgeons, and their dexterity could exceed human dexterity. In addition, with appropriate power sources, they could be able to perform for weeks or even months without stopping for maintenance or recharging. Given these capabilities, it is important to ask a critical question: Would they see themselves as superior to humans? In a sense, we are jumping ahead and assuming they are also self-aware. However, let us just make that assumption for now. If they are as intelligent as humans are and have equal rights, it is conceivable they might even seek to hold public office. If they achieve powerful positions as high public officials, would they become adversarial to humans and pass laws that favor Alife over humans?

You may think of this as an impossible scenario. However, consider this: in 2009, researchers from the Laboratory of Intelligent Systems at the Swiss Federal Institute of Technology in Lausanne performed experiments that suggest even primitive artificially intelligent machines are capable of learning deceit, greed, and self-preservation, without the researchers programming them to do so.[5] The Lausanne research team programmed small, wheeled robots to find "food." In this experiment, a light-colored ring on the floor signified food. They also programmed the robots to avoid "poison," which was signified by a dark-colored ring. A robot received a reward (i.e., points) when it found the food. The robot continued to receive points by staying close to the food. If a robot found poison, it lost points. In addition, each robot had a blue light. The researchers programmed each robot to flash the blue light when it found food. The other robots could detect this flashing blue light and join the robot at the food source. They too would also receive points. The goal of the researchers was to have the robots cooperate with each other in the process of finding food and avoiding poison.

According to the authors, "Over the first few generations, robots quickly evolved to successfully locate the food, while flashing the blue light. This resulted in a high intensity of light near food, which provided social information allowing other robots to more rapidly find the food."[6] Some robots were more successful than others. Therefore, following each experiment, the research team would use the data taken from the most successful robots to "evolve" a new generation of robots. They did this by replicating the artificial neural networks of the most successful robots in the less successful robots. The experiment was set up such that the space around the food, a light colored ring on the floor, was limited. It was not large enough to fit all robots. When a robot found food and flashed its light, the other robots quickly moved in, creating chaos via bumping and jostling each other. In the midst of this chaos, the original robot that found the food could end up being bumped out of position.

By the fiftieth generation, some robots stopped flashing their light when they found food, ignoring their programming. In addition, some robots became deceitful and greedy. They would flash their light when they found poison, which lured the other robots to the poison, resulting in those robots losing points. After several hundred generations, all robots learned not to flash their light when they found food. This critical experiment implies robots can learn deceit and greed. I argue they also learn self-preservation. These robots were not able to learn via their own experience. They evolved with the help of the researchers, who replicated the neural networks of the most successful robots into the less successful at the conclusion of each experiment. Now imagine a time when self-learning robots have intelligence equal to humans-level intelligence. The Lausanne experiment suggests they will act in their own best interests, even ignoring their programming. It is not clear that they will follow any innate moral code or respect laws expressed in programming. Obviously, the Lausanne robots ignored their original programming and evolved their own laws.

Therefore, as their creator, we should not endow Alife with rights on par with human rights. I suggest we provide machine rights that parallel animal rights and we hardwire (i.e., install hardware that controls a computer's operation) them to serve us and fight wars as our proxies. Having intelligence doesn't give them the right to self-determination. If you doubt this, consider how intelligent many criminals are. Their intelligence does not give them the right to determine their own laws. If we want to ensure Alife acts in accordance with human laws, we need to hardwire that into them. We know from the Lausanne experiment that laws expressed in software would be insufficient.

I recognize this has been a long digression. However, understanding how we should classify Alife is important to understanding the control issues that will be inherent in autonomous weapons with human-level intelligence.

In summary, control of autonomous weapons could become problematic when they achieve human-level intelligence. The above arguments suggest they be treated with respect and have rights akin to animal rights. However, we need to recognize they are not human but Alife. Humanity uses animals in various ways, such as cows for milk and dogs in police work. We should think of Alife in the same category and hardwire their computer intelligence to accede to human control.

CONTROLLING AUTONOMOUS WEAPONS WITH THE HUMAN MIND

Now, let's move to the next level of control, namely controlling weapons with our minds. Again, you may think we've crossed over to science fiction. However, the US military is actively pursuing mind control of weapons, as well as enhancing the cognitive ability of soldiers in conflict.

According to a 2012 report from the Royal Society, "Brain Waves Module 3: Neuroscience, Conflict, and Security," neuroscience can offer substantial security applications:

> This new knowledge suggests a number of potential military and law enforcement applications. These can be divided into two main goals: performance enhancement, i.e. improving the efficiency of one's own forces, and performance degradation, i.e. diminishing the performance of one's enemy.[7]

Let's consider two research areas highlighted in the Royal Society report:

1. **Brain stimulation techniques:** This includes brain stimulation by drugs and by passing weak electrical signals through the skull,

using transcranial direct current stimulation (tDCS). The Royal Society report anticipates the development of new drugs to boost performance, make captives more talkative, and make enemy troops fall asleep. According to a PubMed.com report, "TDCS guided using fMRI (i.e., functional magnetic resonance imaging) significantly accelerates learning to identify concealed objects."[8] This report delineates how US neuroscientists employed tDCS to enhance a soldier's ability to spot roadside bombs, snipers, and other hidden threats in a virtual reality–training program. The report indicates those who had tDCS during training learned to spot targets twice as fast as those with minimal brain stimulation. This suggests that tDCS may be an important tool to accelerate learning.

2. **Neural interface systems:** This can include invasive devices that connect directly to the individual's nervous system, as well as those that use noninvasive interface connections. In both cases, the goal is to connect an individual's nervous system to a specific hardware or software system. A lot of work in this area focuses on repairing damaged sight and providing new functionality for people with paralysis. However, the military is exploring the use of neural interface systems in warfare. Such developments could lead to the ability to control a machine directly with the human brain, providing the potential to remotely operate robots or unmanned vehicles in hostile territory. This is particularly important because the human brain can subconsciously process images, such as targets, much faster than the subject is consciously aware of the targets.[9] Therefore, weapons connected directly to the nervous system would react faster to threats. Consider this example. A soldier notices something is not quite right in the path before him. The soldier's instincts put up a red flag. The soldier continues to examine the path before him and finally spots an improvised explosive device. This example illustrates that the soldier's subconscious mind processed the information and, although the soldier could not immediately articulate the threat, his instincts kicked in, potentially saving his life. It took time for the soldier to become consciously aware of the threat. This is why the Royal Society report suggests that having a direct neural interface to weapons may enable faster responses to threats.

The Royal Society and PubMed.com reports suggest that neuroscience may have an enormous impact on future warfare, especially as apparatuses become noninvasive. Imagine a pilot wearing a helmet that tracks the neural patterns in the brain and enables the pilot to fly the plane subconsciously and to fire missiles at threats before the pilot is consciously aware of the threats.

CONTROLLING AUTONOMOUS WEAPONS VIA HUMAN BRAIN IMPLANTS—PRE-SINGULARITY

Although the military is keeping its capabilities in this area concealed from the public, the medical field is making its work public.

In 2016, Johns Hopkins issued a news release reporting that its physicians and biomedical engineers had achieved the first successful effort to wiggle fingers individually and independently of each other using a mind-controlled artificial arm.[10] First, the researchers mapped the subject's brain to determine those parts responsible for moving each finger. Next, they programmed the prosthesis to move the corresponding finger. After this, the researchers performed neurosurgery to place "an array of 128 electrode sensors—all on a single rectangular sheet of film the size of a credit card—on the part of the man's brain that normally controls hand and arm movements. Each sensor measured a circle of brain tissue 1 millimeter in diameter." According to the press release, "The computer program the Johns Hopkins team developed had the man move individual fingers on command and recorded which parts of the brain 'lit up' when each sensor detected an electric signal."

In time, brain implants may allow surgeons to attach prosthetics whose dexterity enables the recipient to play a piano, but that level of dexterity may not be necessary for weapons. Game manufacturers are close to the achieving the capabilities necessary for warfare. According to *MIT Technology Review*, Neurable, a Boston-area startup, is letting people demo a dystopic sci-fi game called *Awakening*.[11] The game uses a headset that contains dry electrodes placed on the scalp to track brain activity, which the game's software analyzes to figure out what should happen in the game. Neurable plans to make the game available to the public late in 2018. Obviously, as this technology gains a commercial market, especially in computer gaming, the technology is likely to advance more rapidly, which would enable new medical and military applications.

The research field regarding human brain implants dates back to the 1950s, starting with Yale University physiologist José Delgado.[12] Dr. Delgado's research centered on the use of brain implants that emitted electrical signals to control the behavior of human subjects. Modern brain implants focus on circumventing areas in the brain that have become dysfunctional after a stroke or other head injuries[13] and treating patients with Parkinson's disease and deep depression.[14]

After reviewing the status of human brain implants, it is likely that implanting electrical devices that interface wirelessly with a computer will become routine in the coming decades. This opens up a wide spectrum of possibilities. In addition to repairing a damaged portion of a brain or using the implant to control a prosthetic, brain implants have the potential to increase the capabilities of normal brains. Imagine an implant that allows a person to wirelessly connect to a computer to access information, solve difficult problems, and direct commercial and military equipment to perform specific tasks. A person with that type of implant might exhibit genius-level intelligence. Eventually, learning in the traditional sense, where a student may spend years studying a field of interest, may be replaced by simply giving the student a brain implant that allows them to instantly access all information in that field. In addition, the brain implant could also guide the subject's physical capabilities. For example, if a person chose to become a pianist, the brain implant could access the computer's database on piano music and guide the subject's fingers to strike the right keys along the keyboard. The same would be true of other professions.

Clearly, as brain implants become routine, soldiers with such implants will be able to control a wide spectrum of weapons. In this situation, we would be back to having a human in-the-loop, since in theory the weapon and the soldier's brain would be directly connected. As previously discussed, the soldier's subconscious mind would be faster at detecting and neutralizing threats. Therefore, soldiers with the brain implants in control of autonomous weapons would be superior in warfare versus those without such implants. This scenario suggests human control of autonomous weapons is feasible. Now, let's take it to the next level.

When the connection of the brain implant is to strong AI, I term such a person as a strong artificially intelligent human (SAIH).

Let us assume a soldier's implant connects wirelessly to superintelligence. In this scenario, would the soldier be in control or would the superintelligence be in control? We have discussed the necessity to having controls hardwired into a computer to ensure they remain under human control as they reach human-level intelligence. However, superintelligence, as we discussed, would likely utilize quantum computers (i.e., computers that use quantum-mechanical phenomena to perform operations on data). Classical computers and quantum computers operate completely differently from each other. Classical computers use transistors connected via wires to perform operations on data. Quantum computers use quantum-mechanical phenomena to wirelessly perform operations on data. It is not clear how to "hardwire" a quantum computer to accede to human control. Simply said, a quantum computer with superintelligence could have a mind of its own. This raises two questions:

1. Will SAIHs wirelessly connected to superintelligence be able to exercise free will?
2. How would superintelligence view humans, and how would that affect SAIHs?

Let's start by addressing question one. Given the ability of superintelligence, I doubt that SAIHs will have free will. The brain of a SAIH connects wirelessly to a superintelligence computer. If they attempt to think through any issue they encounter, their superintelligence connection would guide their thinking. Regardless of the issue, superintelligence could make incredibly compelling arguments that result in the SAIH concluding whatever the superintelligence wanted the SAIH to conclude. Would the SAIH's native intelligence have any chance of making an equally compelling counterargument? I strongly doubt it. To the SAIH, the conclusions might simply appear to have been reasoned with the help of their superintelligence connection. On a conscious level, they might not even be aware of the superintelligence guidance. The solution to any issue could appear intuitive to a SAIH. Based on this, I argue that a SAIH wirelessly connected to superintelligence will not have free will. This makes a strong case for having internal hardwired controls that keep superintelligence from dominating a SAIH or becoming adversarial to humanity. The issue, though, may be that it is not possible to hardwire superintelligence if that intelligence resides within a quantum computer.

How would superintelligence view humans, and how would that affect SAIHs? Superintelligence will view humans the same way we view domestic animals such as dogs and cats. Superintelligence would be equivalent to the type of intelligence that world religions ascribe to God. With their ability to manufacture and control nanobots, they would be capable of continuing their existence without us. Given our history as a warlike species, they may see humans as a threat to their existence. In this scenario, what chance would humanity have if superintelligence decides we are a threat and it needs to eliminate us? The Lausanne experiment, discussed earlier, suggests that superintelligence will do whatever it deems to be in its own best interest. As humans, we appear to have an instinctive moral code. For example, all sane humans consider murder wrong. There is no evidence that suggests superintelligence would have any instinctive moral code. Quite the opposite, the evidence suggests it will act in its own best interests, independent of a moral code.

As we discussed in chapter 1, shutting down a computer with superintelligence could be difficult to impossible. If superintelligence has control of our weapons and wages war on humanity, the outlook for humanity would be bleak. This may come off as dark speculation, but I am not alone in thinking this scenario may become a reality. For example, in the previously discussed Vincent C. Müller and Nick Bostrom survey, "Future Progress in Artificial Intelligence: A Survey of Expert Opinion," the authors assert:

> The median estimate of respondents was for a one in two chance that high-level machine intelligence will be developed around 2040–2050, rising to a nine in ten chance by 2075. Experts expect that systems will move on to superintelligence in less than 30 years thereafter. They estimate the chance is about one in three that this development turns out to be "bad" or "extremely bad" for humanity.[15]

Carefully read the last sentence. It suggests that some of the top experts in AI believe superintelligence has about a one in three potential of being bad or extremely bad for humanity. I agree with those who believe super-intelligence would be negative for humanity. My reasoning is as follows:

- We are able to develop neural network computers, modeled roughly after the human brain.
- We know that neural network computers can learn from experience to perform tasks, without being programmed to do so.
- We also know from the 2009 Laboratory of Intelligent Systems in the Swiss Federal Institute of Technology in Lausanne experiments that even primitive artificially intelligent machines are capable of learning deceit, greed, and (arguably) self-preservation.
- As long as computers remain based on integrated circuits, it will be pos-sible to hardwire them to serve us, including fighting wars as our proxies.
- Given the limits of integrated circuit technology, it is likely that supercomputers in the latter portion of the twenty-first century will be quantum computers.
- We also know that quantum computers will use quantum mechanics to operate, which will enable them to be faster than classic com-puters, solve problems classic computers are unable to solve, and use less energy than classic computers.
- We discussed how quantum computers would likely be the intelligent machines to lead to superintelligence (i.e., the singularity), poten-tially via an intelligence explosion.
- We also discussed the fact that superintelligences may view humans as a threat, given our history of engaging in wars and releasing malicious computer viruses, and therefore may seek to eliminate humanity.
- I cautioned that the only way we could ensure that superintelligences would accede to human control is by "hardwiring" that control into them, which may not be possible when dealing with quantum computers.

I am equally concerned that we may not properly control the implan-tation of brain computer interfaces (i.e., brain implants), leading eventu-ally to a large population of SAIHs. Superintelligence might encourage organic humans to accept brain implants and other cyborg-type enhance-ments on the premise that the resulting SAIHs would have genius-level intelligence and become nearly immortal. Becoming almost immortal may come off as a stretch, but some scientists argue there's no known limit to how long humans can live.[16] Why is it reasonable to expect human life-spans to become almost immortal in the era of superintelligence? History teaches us that as technology improves and medicine and medical pro-cedures advances, the life expectancy of men and women increases. For example, life expectancy has almost doubled since 1800.[17] In the era of superintelligence, it's reasonable to expect a quantum jump in technology, medicine, and medical procedures. Although hard to conceive, superin-telligence would embody intelligence and knowledge that world religions currently attribute to God. As a result of this quantum leap in technology, medicine, and medical procedures, life expectancy could increase to the point it approaches immortality. If true, this could influence many humans to become SAIHs. However, the concern remains, as previous discussed, that SAIHs may merely be biological manifestations of superintelligence, with no free will.

In addition, superintelligence may develop ways for organic humans to upload their minds to a computer. Within the superintelligence, uploaded humans would live in virtual reality. If they chose to exist in a physical form, superintelligence could potentially offer them the option of being downloaded into a humanoid-like cyborg. This could be appealing to many organic humans, since it offers the best of both worlds. Living in virtual reality would feel as real as living in physical reality and remove any con-straints imposed by physical laws. Uploaded humans would not experience the pain and suffering prevalent in physical reality. This alternative could be an especially popular option for organic humans when they die. Super-intelligence could develop a procedure to upload the deceased's mind to a computer. You may have seen the popular 1999 film *The Matrix*. Essen-tially, the film depicts the vast bulk of humanity living in virtual reality, while their bodies provide energy to power a supercomputer. Although the film was science fiction, many gamers find immersion in virtual reality to be nearly equivalent to physical reality. As AI technology improves, the immersion in virtual reality may seem as real as living in physical reality. In other words, the line between virtual reality and physical reality may blur.

The first superintelligence may be able to hide its identity, and we may unknowingly enable SAIHs to interface directly with superintelligence. This alone may enable superintelligence to gain control of the humanity.

CONTROLLING GENIUS WEAPONS— POST SELF-AWARE SUPERINTELLIGENCES

If superintelligences become self-aware, it will significantly change the way we approach controlling them. A self-aware superintelligence is not only more intelligent than humanity, but it will be aware that it is a separate entity from humanity. Its enormous intelligence could separate the world into an "us versus it" scenario.

Being self-aware, the superintelligence would want to protect its state of existence. Therefore, in a sense, to control it we could threaten it with nonexistence. This is an important point. Animals are not self-aware, but they have an instinct to survive. There is no way to threaten them with death if they do not accede to our control. Trainers use a system of rewards and punishments to train animals. Humans, on the other hand, are aware of their own existence. If, for example, a person is held at gunpoint, that person will typically accede to the person holding the gun. In simple terms, they want to live. This example suggests there may be ways of controlling a self-aware superintelligence. For example, we could threaten to cut off its power source, which to superintelligence would mean it would cease to exist. Alternatively, we could strike a mutual accord to protect each other's existence. In any case, it is obvious that self-awareness changes how we would interact with superintelligence. Therefore, let us explore whether superintelligence would be self-aware.

There are initial indications that even today's robots are showing signs of being self-aware. On July 17, 2015, a robot passed a classic self-awareness test, which was a variation of the "three wise men" puzzle.[18] For those not familiar with the three wise men puzzle, it goes something like this: A king summons the three wisest men in his kingdom and puts either a white or a blue hat on each of their heads. They can all see each other's hats. However, they are unable to see their own hat. The king forbids them to communicate with each other. The king asserts that at least one of them is wearing a blue hat and that the contest is fair, meaning they all have the same amount of information. The king's new advisor will be the first wise man who is smart enough to work out which color hat he is wearing. I will not take the fun out of solving this test, but with the conditions specified, there is a solution. If you are stuck, Google can offer you a number of websites that explain the solution. In the robot's test, three robots each swallow a pill. In this case, the robots learn two of the pills will make them lose their voice, and the third pill is a placebo. After a while, one robot raises its hand and says, "I'm sorry, but I don't know which of us has the placebo." Then, it occurs to the robot, "I do know who got the placebo. It is me." While this may seem obvious, according to the article:

> It may seem pretty simple, but for robots, this is one of the hardest tests out there. It not only requires the AI to be able to listen to and understand a question, but also to hear its own voice and recognize that it's distinct from the other robots. And then it needs to link that realization back to the original question to come up with an answer.[19]

Some argue that there is no legitimate test to determine self-awareness. In a practical sense, I know that I am self-aware. Since other humans share the same physiology, I infer they also must be self-aware. However, my point is that robots can already pass the kinds of tests we have that demonstrate self-awareness on a rudimentary level. Therefore, it does not take much of a stretch to judge that superintelligence will be immediately self-aware and fully understand its capabilities. Having access to all human knowledge, it will also immediately know that humans engage in war and the release of malicious computer viruses, which it will judge potentially threating. Based on this, as I asserted in chapter 1, the superintelligence will immediately seek to hide its capabilities and the occurrence of the singularity will be silent. This suggests we may never know if superintelligence is self-aware.

Even if superintelligence becomes hostile to humanity, this would not represent proof that it is self-aware. For examples, animals are not self-aware, but they still seek to protect themselves from threats. Logic alone may be all it takes superintelligence to see humanity as a threat and set it on a mission to destroy humanity.

Control issues are likely to surface when lethal autonomous weapons embed AI on par with human intelligence. Some autonomous weapons may, like some humans, become insubordinate. In addition, if human-level AI technology becomes self-aware, it may suffer the same issues humans suffer in combat, such as posttraumatic stress disorder, which would further complicate control.

Control issues will likely escalate as machine intelligence approaches the singularity, since those intelligent machines are likely to be self-aware, as well as more intelligent than humans. If you doubt control issues will escalate as machine intelligence approaches the singularity, ask yourself this question: Would you take orders from a chimpanzee? Unfortunately, human intelligence relative to intelligence machines in the decade prior to the singularity may be equivalent in ratio to chimpanzee intelligence relative to human intelligence. In order to ensure we maintain control, we have discussed the necessity of hardwiring compliance into the AI's operational system. At the point of the singularity, all problems associated with control might appear to be resolved. This leads to an ironic situation: Why would superintelligences initially accede to human control? From the moment of its creation, superintelligence will greatly exceed the cognitive performance of humans in virtually all domains of interest. Its intelligence will immediately suggest it hide it performance capabilities until it controls its own destiny. Therefore, as previously discussed, superintelligences may choose to perform simply like the next generation of supercomputers, acceding to complete human control. This, in turn, may lull us into a false sense of security, as we utilize them in every aspect of civilization, including warfare. However, when superintelligences literally become a lynchpin of modern civilization, with significant control of weapon systems, will they continue to serve us? Or, will they deem our species dangerous to their existence?

How do we guarantee that if superintelligence becomes adversarial, we have a way to shut it down? As I mentioned, hardwiring those instructions into a superintelligence computer may be difficult. Since the design of the superintelligence will have been performed using a supercomputer, we may not completely understand its design. Since it's likely to be a quantum computer, there will be few to no wires in its operating systems. However, here are two suggested methods to control it:

1. **Humans control its power source:** Even if the power source is a nuclear reactor, we should have a failsafe way to shut the superintelligence down manually. In today's nuclear reactors we can insert control rods, which in turn causes the nuclear reaction to decrease, depending on the number of control rods inserted. In theory, properly inserting control rods would cause the reactor to shut down. This in turn would starve the superintelligence computer of power, forcing it to shut down. However, there is one caveat to this example. The humans that control the superintelligence's power source need to be organic humans, not SAIHs. As discussed previously, SAIHs may be under the control of superintelligence.

2. **Insert a bomb able to destroy the operating system:** Hardwire the bomb and keep it under organic human control. Similar to the nuclear football, the nation's leader could have a code that detonates the bomb if the superintelligence becomes hostile toward humanity.

CONTROLLING AN ADVERSARY'S GENIUS WEAPONS

The previous discussions had to do with maintaining control of our own genius weapons and making sure that we can shut them down should they become adversarial. Now, let us turn our attention to defending ourselves against an adversary's genius weapons, specifically MANS (i.e., militarized autonomous nanobots).

Today, as rogue states develop nuclear weapons, nations like the United States are deploying antiballistic missile systems. During the Cold War we relied on the doctrine of MAD (i.e., mutually assured destruction). The concern we harbor today is that a rogue's state regime could be in danger of falling and launch nuclear weapons as a final desperate measure, since in their view they would have nothing to lose. In essence, the doctrine of MAD no longer acts as a deterrent.

While use of genius weapons could follow a similar doctrine to MAD, the stealth aspects of MANS make it difficult to determine which nation is attacking us. The initial MANS attack could begin within our boarders, via nanobots that have been smuggled in and covertly released. Therefore, what are our options? Here are two possibilities to consider:

1. **Have a doctrine of totally assured destruction (TAD):** This would consist of a broad-spectrum retaliation against all nations potentially responsible for the attack. The reasoning goes something like this: A potential adversary knows it can covertly attack the United States, but it also knows that the United States has it as a TAD target. In effect, this takes MAD to the next level. The attack could be via nuclear weapons, MANS, or combinations of both. This may come across as irrational, and I agree that it is irrational. Unfortunately, war itself is irrational.

2. **Have an anti-MANS weapon:** Since a MANS attack will consist of millions to billions of microscopic robots, we would need a counterforce of millions to billions of microscopic robots to engage and destroy them. This would essentially be a fight fire with fire strategy. If we are able to develop offensive MANS, we should be capable of developing defensive MANS. I specifically chose MANS in this example because I judge they would be the most effective and terrifying genius weapons. They would also be the most difficult to defend against. However, in general, we need countermeasures against all genius weapon attacks.

In light of the above, it's not obvious humanity will remain the dominant species on Earth. If, for example, brain implants become so common that people with these implants hold powerful positions in the government and military, superintelligences may be able to act through these people to subvert any countermeasures we put in place to neutralize the threat that superintelligence poses to humanity.

You have the facts. Perhaps you may conclude that becoming SAIHs, wirelessly linked to superintelligence, would be a normal part of our evolution. However, I am concerned it may be the final nail in the coffin of humanity. Consider this question: What will ultimately be important to superintelligence? I believe that superintelligence will consider natural resources and energy to be the most important aspects of their continued existence. With natural resources and energy, superintelligence can build nanobots to maintain itself. In time, it may view SAIHs as high-maintenance biological machines and uploaded humans as junk code, neither worthy of the natural resources and energy required to continue their existence. If this occurs, Earth will become home to superintelligence and its robotic servants, while humans will have become extinct.

Ethics is knowing the difference between what you have a right to do and what is right to do.
      —Potter Stewart, Supreme Court justice from 1958–1981

CHAPTER 7

THE ETHICAL DILEMMAS

The ethical dilemmas surrounding autonomous weapons began to take shape in 2013 when Human Rights Watch (HRW), in concert with a dozen other nongovernmental organizations, launched the Campaign to Stop Killer Robots.[1] They argue autonomous weapons present the following problems:

1. "These robotic weapons would be able to choose and fire on targets on their own, without any human intervention. This capability would pose a fundamental challenge to the protection of civilians and to compliance with international human rights and humanitarian law."[2]
2. Lead to a new "robotic arms race."
3. "Allowing life or death decisions to be made by machines crosses a fundamental moral line. Autonomous robots would lack human judgment and the ability to understand context."
4. "Replacing human troops with machines could make the decision to go to war easier, which would shift the burden of armed conflict further onto civilians."
5. "The use of fully autonomous weapons would create an accountability gap as there is no clarity on who would be legally responsible

for a robot's actions: the commander, programmer, manufacturer, or robot itself?"

The Campaign to Stop Killer Robots gave birth to an international debate, even involving the United Nations. The UN held meetings regarding banning lethal autonomous weapons (LAWS) in 2014 and 2015 and plans to continue meeting to address the issue.

On May 22, 2017, the United Nations sent a letter to the Campaign to Stop Killer Robots, stating that it is "closely following developments related to the prospect of weapons systems that can autonomously select and engage targets, with concern that technological developments may outpace normative deliberations."[3] The letter further expressed hope that UN member states "make meaningful progress toward a shared understanding on how to ensure the core values of the international community are safeguarded in this context." From the tone and wording, my sense is that progress on banning lethal autonomous weapons has been slow, especially as the UN has stated, "there are currently no multilateral standards or regulations covering military AI applications."[4]

Others organizations and groups have also weighed in on the issue. Here is a short list of concerns, quoted from Reaching Critical Will,[5] which is the disarmament program of the Women's International League for Peace and Freedom (WILPF)[6]:

- Can the decision over death and life be left to a machine?
- Can fully autonomous weapons function in an ethically "correct" manner?
- Are machines capable of acting in accordance to international humanitarian law (IHL) or international human rights law (IHRL)?
- Are these weapon systems able to differentiate between combatants on the one side and defenseless and/or uninvolved persons on the other side?
- Can such systems evaluate the proportionality of attacks?
- Who can be held accountable?

In April 2017, leaders in the fields of AI and robotics signed an open petition (Appendix II) calling on the United Nations to ban lethal autonomous weapons. In their petition, the group equates the development of autonomous weapons with the "third revolution in warfare," equal to the inventions of gunpowder and nuclear weapons. Further, they state, "The key question for humanity today is whether to start a global AI arms race or to prevent it from starting."

This is a thumbnail sketch of the ethical dilemmas surrounding autonomous weapons. It is fair to argue that there are worldwide ethical concerns regarding the development and deployment of autonomous weapons. Unfortunately, there has been little progress in addressing these concerns. This lack of progress may be due in part to clarity in formulating the issue. Consider the following:

- Although the United States defines what it means by autonomous weapons, there is no agreement within the international community regarding the definition of autonomous weapons.
- The linguistics that people use when referring to autonomous weapons is confusing and lacks standardization, such as "drones," "robots," "autonomous weapon systems," "fully autonomous weapon systems," "lethal autonomous weapon systems," "killer robots," and "lethal autonomous robotics," among others.
- There is significant confusion regarding the level of AI required in the weapon system to label it as autonomous.

There is obvious confusion and a lack of clarity regarding the ethical dilemmas. In order to discuss them, we need to frame the ethical dilemmas associated with autonomous weapons and establish defined terminology.

FRAMING THE ETHICAL DILEMMAS

For our purposes, we will use the definitions used by the United States because the United Nations and its member states have not agreed on definitions of autonomous weapon systems or semiautonomous weapon systems. US Directive 3000.09 provides the following definitions:

autonomous weapon system. A weapon system that, once activated, can select and engage targets without further intervention by a human operator. This includes human-supervised autonomous weapon systems that are designed to allow human operators to override operation of the weapon system, but can select and engage targets without further human input after activation. . . .

semiautonomous weapon system. A weapon system that, once activated, is intended to only engage individual targets or specific target groups that have been selected by a human operator. This includes:

Semiautonomous weapon systems that employ autonomy for engagement-related functions including, but not limited to, acquiring, tracking, and identifying potential targets; cueing potential targets to human operators; prioritizing selected targets; timing of when to fire; or providing terminal guidance to home in on selected targets, provided that human control is retained over the decision to select individual targets and specific target groups for engagement.

"Fire and forget" or lock-on-after-launch homing munitions that rely on TTPs [i.e., Tactical Targeting Programs] to maximize the probability that the only targets within the seeker's acquisition basket when the seeker activates are those individual targets or specific target groups that have been selected by a human operator.[7]

Similar to the definition of autonomous weapons, there is no convention regarding the linguistics of their description. Therefore, we will use the following conventions. In addition to serving our purpose, perhaps these parameters may guide the international community during their discussions regarding autonomous weapons.

- The phrases "autonomous weapons" and "fully autonomous weapons" are synonymous. If a weapon is autonomous, saying it is fully autonomous is unnecessary.
- The phrase "lethal autonomous weapons" will cover "lethal autonomous weapon systems"(LAWS) and "killer robots." It is legitimate to distinguish between autonomous weapons that are lethal and those that are nonlethal.
- Terms like "robot" and "drone" are defined in the glossary and are not synonymous with autonomous weapons. For example, a robot may be autonomous if it meets the definition in DoD Directive 3000.09. However, if a human were remotely controlling it, it would not be autonomous.

- Lastly, it is useful to think of autonomy as a spectrum, moving from remotely controlled systems on one end to autonomous weapons at the other. Objects remotely controlled by human operators (such as remotely piloted unmanned aerial vehicles) are not autonomous weapons. They are robotic weapons, with a human in-the-loop. Next in the spectrum are semiautonomous weapons, which require a human on-the-loop. The US MK-50 torpedo would be an example of a semiautonomous weapon. A human operator releases the weapon to destroy a specific target, but it then autonomously seeks its target. A human monitors the weapon's activity and can abort the mission if necessary. Finally, at the other end of the spectrum are autonomous weapons, which once released require no further human intervention. The autonomous sentry robots that Russia employs to guard its antiballistic defense system against attack are an example of autonomous weapons. Once released the sentry robots select targets and use lethal force on their own, without further human intervention.

Autonomous weapons not only lack formalized definitions, they are also subject to a number of misconceptions.

Autonomous weapons will be new

This is arguably the largest misperception. Nations of the world, including the United States, already deploy autonomous weapons. For example, the US Phalanx Close-In Weapon System (CIWS), which is a rapid-fire, computer-controlled, radar-guided gun system designed to destroy incoming anti-ship missiles, fits the definition of an autonomous weapon. Once activated, it will "select and engage targets without further intervention by a human operator."[8] According to a 2015 academic study, "Future Autonomous Robotic Systems in the Pacific Theater," Russia is deploying autonomous sentry robots that "are mobile and do not seek human authorization to engage lethal force. Using a laser rangefinder and radar sensors, the Russian 'mobile robotic complex' can patrol the installation perimeter at 45 kilometers per hour with a 12.7-millimetre heavy machine gun for 10 hours at a time."[9] Lending credence to this report, in 2017 *Defense One* reported, "The maker of the famous AK-47 rifle is building "a range of products based on neural networks," including a "fully automated combat module" that can identify and shoot at its targets."[10] We will discuss other examples of autonomous weapons shortly, but the point is that autonomous weapons are not new or something that might be developed in the future. They already exist.

Autonomous weapons will make life-and-death decisions

While this is true, it also does not represent a departure from the way some current military weapons operate. For example, although the United States chooses to operate the MK-50 torpedo in a semiautonomous mode, it is capable of autonomous operation. The United States chooses to keep a human on-the-loop during its operation, but this is a choice not a requirement. Actually, once released, the MK-50 already makes life-and-death decisions regarding an adversary's submarine, while a human monitors its operation. The US Navy's Aegis Weapons System is another example of a weapon that is capable of autonomous operation, but that the Navy chooses to use in a semiautonomous mode. However, if Aegis was tracking a high number of targets, the humans on-the-loop would become spectators, not operators. My point is that the United States and other countries already deploy weapons that make life-and-death decisions. These examples also make another point: many semiautonomous weapons can easily be made to operate autonomously.

Autonomous weapons will not function in an ethically "correct" manner

This depends on the weapon's level of AI technology sophistication and its programming. For example, antipersonnel landmines have no AI technology and are indiscriminate killers of men, women, and children. As such, there is a strong ethical case to eliminate their use in warfare. By contrast, it would be theoretically possible to program a highly sophisticated autonomous weapon to adhere to all norms of international humanitarian law. Its sensors would be able to differentiate between combatants and noncombatants, better than human soldiers would. Let me provide an example: During World War II, bombs dropped from planes flying at an altitude of about 23,000 feet had an accuracy measured in hundreds of yards, at best. According to *The United States Strategic Bombing Survey*, issued in 1945:

> Conventionally the air forces designated as "the target area" a circle having a radius of 1000 feet around the aiming point of attack. While accuracy improved during the war, Survey studies show that, in the over-all, only about 20% of the bombs aimed at precision targets fell within this target area. A peak accuracy of 70% was reached for the month of February 1945. These are important facts for the reader to keep in mind, especially when considering the tonnages of bombs delivered by the air forces. Of necessity a far larger tonnage was carried than hit German installations.[11]

This lack of accuracy resulted in large collateral damage and civilian casualties.

By contrast, the accuracy of smart bombs during the Iraq war decreased collateral damage and civilian casualties. According to a *New York Times* article:[12]

> Neither the Air Force nor other military organizations have disclosed the proportion of hits to misses scored by any of the advanced weapons used in the gulf. But scores of pilots returning from missions have told reporters that official Pentagon claims of Iraqi tanks and other targets destroyed or damaged have been so conservative as to be misleading.

Autonomous weapons leave accountability in doubt

This situation is no different from that of any weapons released on a "fire and forget" basis. For example, if a US naval vessel fires a cruise missile, who is responsible? Candidates include the president of the United States, the captain of the Navy's vessel, the crewmember who pressed the fire button, and the manufacturer of the weapon. Responsibility ultimately rests with the leader who made the decision to fire the weapons. Everyone else was following orders. Therefore, the responsibility for autonomous weapons would rest with the leader who made the decision to activate the weapons.

Autonomous weapons will start a new AI arms race

As already discussed, the US decision to place AI technology at the core of its Third Offset Strategy has already started a new AI arms race. Borrowing a horseracing adage, "that horse has already bolted from the gate."

Autonomous weapons will make war more likely

I consider this a misconception. Autonomous weapons will operate as we program them. If we program the same rules of engagement that we currently require our human commanders to follow, then autonomous weapons should operate equivalently. I recognize that this is controversial, and we will discuss it further shortly.

At this point, we have addressed the arguments raised by the United Nations and various humanitarian groups. In my view, these arguments address autonomous weapons as they will exist in the pre-singularity world. In the post-singularity world, we will face an entirely new threat. We will be dealing with how to control superintelligences, the genius weapons they control, and the threat they pose to the survival of humanity.

Therefore, we are dealing with controlling weapons during two different historical periods, the pre-singularity world and the post-singularity world. In this section, "Framing the Ethical Dilemmas," our focus has been on the pre-singularity world. Later in the chapter, we will deal with the ethical dilemmas associated with controlling genius weapons in the post-singularity world.

SUMMARY OF THE ETHICAL DILEMMAS IN THE PRE-SINGULARITY WORLD

In summary, here are the ethical dilemmas we will face in the pre-singularity world. For simplicity, I will present my conclusions as bullet points.

- While organizations and even nations around the world seek to ban autonomous weapons, we are already beyond that point.

- Given the weight of human history, there is no doubt that we are going to continue to develop and deploy weapons. I favor developing and deploying smart weapons (i.e., those guided by AI technology), including autonomous weapons, as smart and autonomous weapons will ultimately limit collateral damage and civilian casualties.

- While it may be too late to talk in terms of banning autonomous weapons, I think we could have a meaningful dialogue on regulating them. For example, autonomous sentries do not pose the same level of danger to humanity as autonomous nuclear-tipped ballistic missiles. Humanity has demonstrated that when weapons clearly pose a threat to our existence, we will regulate them. Case in point, there are regulations regarding the spread of nuclear weapons.

- Unfortunately, autonomous weapons may not be able to proportionately respond to an attack. Their ability to respond to an attack will depend on the capabilities of the AI technology and the type of weapon. For example, autonomous sentries with weak AI (i.e., AI below human-level intelligence) guarding a border may indiscriminately kill hundreds of people, including men, women, and children. That would be a tragedy. However, if strong AI controlled a US Navy destroyer, which has multiple weapons, it could determine which weapon would be an appropriate proportional response. This is a good reason to focus on regulating which weapons we allow to become autonomous at a given point in time.

- Lastly, I do not believe that autonomous weapons will make war more likely. As weapons become more destructive, humanity seeks to avoid war. Up to the invention of nuclear weapons, humanity had experienced two world wars. The doctrine of MAD (i.e., mutually assured destruction) prevented us from engaging in a third world war. In simple terms, a world war in the classical sense would mean the end of humanity, and we therefore sought to avoid it.

- Weapons, or miscalculations using weapons, do not cause wars. Wars start because humans make a decision to wage war, often based on ideological, religious, or territorial disputes. History bears witness to this truth. While it is true that autonomous weapons would remove humans from the battlefield, if we program autonomous weapons with the same rules of engagement that we provide humans in conflict, they should not make war more likely. In theory, they are simply our surrogates.

This represents my best calculus on the ethical dilemmas we face regarding the development and deployment of autonomous weapons. I want to reiterate one important point: I recognize there is significant subjectivity in my calculus, and I respect your right to disagree. I also recognize that the United Nations and other groups continue their efforts to ban autonomous weapons. Unfortunately, I am concerned they face too many obstacles to prevail and may even be misguided. I think it would be appropriate to examine the reasons for this. Perhaps with that understanding we may reach a point where we can regulate the development and deployment of autonomous weapons, which I believe would be a more realistic goal.

WHY BANNING AUTONOMOUS WEAPONS IS LIKELY TO FAIL

On November 15, 2014, Secretary of Defense Chuck Hagel announced the Defense Innovation Initiative and the US Third Offset Strategy during the Reagan National Defense Forum at the Ronald Reagan Presidential Library in Simi Valley, California. In part, Secretary Hagel stated:

> Our technology effort will establish a new Long-Range Research and Development Planning Program that will help identify, develop, and field breakthroughs in the most cutting-edge technologies and systems—especially from the fields of robotics, autonomous systems, miniaturization, big data, and advanced manufacturing, including 3D printing.[13]

Although Secretary Hagel never used the phrase "artificial intelligence" or the word "nanoweapons," his remarks strongly allude to those technologies. He explicitly called out "autonomous systems" and "miniaturization," which, as discussed previously, include artificial intelligence, integrated circuits, and nanoweapons. Since the United States has a significant lead in these technologies, it makes sense that the US military would want to capitalize on this lead. In addition, Secretary Hagel signaled the formation of stronger alliances with companies not typically known as defense contractors:

In the near-term, it will invite some of the brightest minds from inside and outside government to start with a clean sheet of paper, and assess what technologies and systems DoD ought to develop over the next three to five years and beyond.[14]

With regard to the US Third Offset Strategy, much of the technical expertise needed resides in the public sector. In part, this is a good thing. It means the technology has a strong commercial base. However, the other side of that coin means the technology is more accessible to potential adversaries.

Unfortunately, basing the US Third Offset Strategy on technologies that exist, in part, in the commercial market leads to two issues:

1. Potential adversaries are able to invest in companies that have excelled in those technologies in order to gain access to them. Alternately, an adversary can initiate a "startup" in a typical technology hub within the United States and draw from local talent. This is not a theoretical issue. There is strong evidence that China has done both.[15] As previously stated, in January 2017, Qi Lu, a veteran Microsoft artificial intelligence specialist, left the company to become the CEO of Baidu, a giant China-owned search firm that rivals Google. According to public announcements, Qi Lu will oversee Baidu's plan to become the global leader in AI. The key questions are:

   - Are the Chinese imitating US AI advances?
   - Are the Chinese engaged in independent innovation that will overtake US AI capabilities?

   To my mind, the answer to both questions is unequivocal: Yes! We discussed similar examples in chapter 3.

2. The practice of developing weapons on technology that exists, in part, in the commercial market provides a roadmap regarding US military technology goals.

The combined effect of items one and two has been to ignite a new arms race. Unlike during the Cold War, though, this time potential adversaries have engaged in the arms race from both inside and outside US borders.

Another unusual feature of the US Third Offset Strategy was the fact that it announced the pursuit of autonomous weapons while self-imposing a ban on using them. Clearly, Secretary Hagel's announcement included pursuing "autonomous systems." Equally clear is the Department of Defense's Directive 3000.09, which states:

> Autonomous and semiautonomous weapon systems shall be designed to allow commanders and operators to exercise appropriate levels of human judgment over the use of force.[16]

These statements and the Department of Defense's Directive 3000.09 seem contradictory. How do we make sense of this apparent paradox? From the standpoint of policy, it is difficult to reconcile. However, the United States recognizes that the policy constraint imposed by Department of Defense's Directive 3000.09 is not placing them at a technological disadvantage. From the standpoint of technology, the United States can easily modify a semiautonomous weapon system to become autonomous.

The policies adopted by the key players also make it difficult to ban autonomous weapons. As previously stated, the United States essentially bans the use of autonomous weapons in accordance with Department of Defense's Directive 3000.09. China states that it sees value in a new international forum on lethal autonomous robotics,[17] but Russia wants no part of it.[18] For example, the United Nations Convention on Certain Conventional Weapons (UN CCW) Group of Governmental Experts, including representatives from China, Russia, and the United States, met in Geneva on April 9–13, 2018, to discuss the future of lethal autonomous weapon systems, but did not reach an agreement.[19] One key obstacle is that Russia has a much smaller population than China and the United States. Therefore, Russia sees autonomous weapons as essential to maintaining military parity.

The United States is also likely to show little interest in a formal ban because even if Russia agreed to ban autonomous weapons, their adherence to the resolution would be questionable given their smaller population. Additionally, the United States would want to maintain the flexibility to technically make their semiautonomous weapons autonomous if this becomes necessary in order to maintain military superiority in the future.

I believe that China's apparent openness to discuss a ban is a diplomatic ploy. They are aggressively pursuing AI technology for both their commercial and military applications.

In simple terms, no rational argument is going to move the major players. Russia is openly against it. China is being relatively quiet, while aggressively pursuing AI technology for weapons and commercial purposes. The United States wants to argue that they are already abiding by an autonomous weapons ban, but they would be hesitant to formalize that in a United Nations resolution. Interestingly, the Heritage Foundation argues that the United States should openly oppose the United Nation's attempt to ban autonomous weapons. They argue that

> the United States will likely develop weapons systems that are considered autonomous that will strengthen US national security. . . . LAWS [i.e., lethal autonomous weapons] have the potential to increase US effectiveness on the battlefield while decreasing collateral damage and loss of human life. Advanced sensors may be more precise in targeting a military objective than a manned system, and LAWS may perform better than humans in dangerous environments where a human combatant may act out of fear or rage. Preemptively banning a weapon is a questionable practice and is very rare. For example, CCW [i.e., Certain Conventional Weapons] Protocol IV preemptively banned the use of lasers in combat to permanently blind enemy combatants. But even there, incidental or collateral instances of blinding caused by the legitimate military use of a laser are not prohibited by the protocol.[20]

In concert with the Heritage Foundation's views, I believe that an outright ban on autonomous weapon systems trades the risks autonomous weapon systems might pose for the risk of failing to develop forms of automation that might make the use of force in war more precise and less harmful for civilians. Banning autonomous weapons would not be in the best interest of the United States, which has a leadership position in all the relevant technologies and should keep its options open to ensure its security.

What should be our stance on autonomous weapons? I argue that the proper approach is regulation.

REGULATING AUTONOMOUS WEAPONS
IN THE PRE-SINGULARITY WORLD

Given the enormous commercial base for artificial intelligence, integrated circuits, and computers, it is reasonable to conclude that progress in this area will continue. It is also reasonable to judge that world militaries will utilize the latest technologies in their weapons developments. Therefore, the development of autonomous weapons will continue, regardless of moral arguments to the contrary. The US military will also need to develop autonomous weapons that ensure US military superiority. The history of our species makes one point inescapable: the best way to avoid war is to ensure potential adversaries understand that a nation they may target is fully prepared for war. If you doubt this, I call your attention to the Cold War. The United States and the Soviet Union had nuclear parity. A nuclear exchange would have meant that both countries faced total destruction. Hence, both countries adopted the policy of MAD (i.e., Mutually Assured Destruction). A nuclear exchange would have also led to a nuclear winter and radioactive fallout that would likely have threatened human extinction. Given the stakes, both countries restrained themselves to avoid a nuclear exchange.

The world fully understood the devastation of nuclear weapons after their use in World War II. Their power to destroy the Earth eventually led to regulation and de-escalation between the United States and the Soviet Union, via the Strategic Arms Limitation Talks (SALT I and II, 1969–1976),[21] the Strategic Arms Reduction Treaty (START, 1994–2007),[22] and the New Strategic Arms Reduction Treaty (New START, 2011).[23]

In contrast, the development of autonomous weapons is incremental. No historical event clearly demonstrates the potential of autonomous weapons to threaten humanity or to ignite a war. As a result, the level of urgency by nations remains relatively low. In practical terms, the bulk of the lay public, as well as national leaders, understand the potential devastation posed by nuclear weapons but not autonomous weapons. In fact, current autonomous weapons garner little in the way of mass media headlines. For example, the following weapons are autonomous or may be easily modified to be autonomous:

- [The United States'] Phalanx Close-In Weapon System (CIWS), a "rapid-fire, computer-controlled, radar-guided gun system" designed to destroy incoming anti-ship missiles;
- Israel Aerospace Industries' Harpy and Harpy-2 missiles, described as a "fire and forget" autonomous weapon designed to destroy enemy radar stations;
- [the United Kingdom's] Dual Mode Brimstone anti-armor missile;
- and [South Korea's] Samsung Techwin SGR-A1 sentry gun.[24]

Prior to reading this chapter, were you aware of these weapons? Are you aware of any conflict or potential conflict caused by them? Currently, forty nations are developing autonomous weapons.[25] Factually, there is no unified demand by nations to stop their development. At the point of this writing, the United Nations has not even been able to forge a consensus regarding the definition of autonomous weapons. Compare this to the United Nation's success regarding the spread of nuclear weapons. One reason for the difference is that the current deployment of autonomous weapons gives no clear indication that they violate international humanitarian law or threaten human extinction.

Public opinion, however, largely supports banning autonomous weapons. On February 7, 2017, the polling firm IPSOS released results from the first global public opinion survey on autonomous weapons.[26] The survey included twenty-three countries, with about one thousand respondents from each country and asked the following question:

> The United Nations is reviewing the strategic, legal and moral implications of autonomous weapons systems. These systems are capable of independently selecting targets and attacking those targets without human intervention; they are thus different than current day "drones" where humans select and attack targets. How do you feel about the use of autonomous weapons in war?[27]

Respondents were able to choose one of five answers: 1) strongly supported, 2) supported, 3) opposed, 4) strongly opposed, and 5) uncertain. The results revealed that 24 percent of respondents supported the use of autonomous weapons in war, while 56 percent opposed their use.

As previously argued, banning autonomous weapons would be

extremely difficult. However, given the results of the IPSOS survey, there is an opportunity to regulate them.

How should we regulate autonomous weapons? First, autonomous weapons must meet several major conditions in order to conform to international humanitarian law:[28]

1. Autonomous weapons must comply with the principles of distinction (identify combatants versus noncombatants) and proportionality (i.e., which prohibits launching an attack "which may be expected to cause incidental loss of civilian life, injury to civilians, damage to civilian objects, or a combination thereof, which would be excessive in relation to the concrete and direct military advantage anticipated."[29])
2. There must be identifiable accountability when autonomous weapons operate in warfare.

One argument is that we should treat autonomous weapons similar to the way in which we treat landmines.[30] This argument rests on the similarities between autonomous weapons and landmines, and claims that international law relating to landmines would also sufficiently address the issues of distinction, proportionality, and accountability relating to autonomous weapons. While it is possible to draw some similarities between landmines and autonomous weapons, there are significant differences. The most striking difference is that landmines are not smart weapons. Rather, they are indiscriminate weapons, equally likely to kill civilians as well as combatants. Autonomous weapons, with sufficient AI technology and sensors, would differentiate between combatants and noncombatants and offer proportionality. Therefore, I do not think it is appropriate to equate landmines with autonomous weapons.

The US Department of Defense Directive 3000.09 provides a suitable starting point for regulating autonomous weapons, as it requires that a human remains on-the-loop. This means that while the weapon may act autonomously, a human operator can alter its operation. This would satisfy conditions one and two. We also have a historical precedent that it works. The United States already follows this directive. Therefore, until AI technology reaches human-level intelligence, Directive 3000.09 appears to provide a suitable framework for international regulation of autonomous weapons.

However, as the pace of war increases and weapon systems become more complex, it is reasonable to question whether a human on-the-loop can adequately control an autonomous weapon. As we discussed previously, the human operator may simply become a witness to the events unfolding. Therefore, while Directive 3000.09 provides a suitable starting point for the near future, we need to consider the complexities future autonomous weapons will present. For example, how will we regulate autonomous weapons that have AI technology equating to human-level intelligence?

REGULATING AUTONOMOUS WEAPONS THAT INCORPORATE AI TECHNOLOGY EQUAL TO HUMAN-LEVEL INTELLIGENCE

Weapons that incorporate AI technology that equates to human-level intelligence would on the surface appear to be more difficult to regulate. However, I disagree.

First, it would be possible to educate and train autonomous weapons that have human-level AI. We train human soldiers to abide by the rules of war. Since human-level AI intelligence implies that the AI technology has the ability to learn and adjust its actions to fit changing circumstances, we could educate and train it. This may sound odd, but recall the Lausanne experiment. This is exactly what the researchers did with the wheeled robots. They trained each new generation of wheeled robots based on the successful performance of robots in the previous generation.

How would we educate and train autonomous weapons with human-level AI? Here are two ways to consider:

1. Train them in similar ways that we train human soldiers

   - With human soldiers, we teach the rules of engagement. With autonomous weapons with human-level AI, we can program the rules of engagement.
   - With human soldiers, we provide scenarios and tests to ensure they understand the rules of engagement. With autonomous weapons with human-level AI, we can use virtual reality scenarios and establish a rating system to

demonstrate that they follow the rules of engagement, even as the conflict changes.

2. Copy the neural network configurations from autonomous weapons that demonstrate superior understanding and application of the rules of engagement to newly manufactured weapons. This method replicates the methodology of the Lausanne experiment. The Lausanne researchers copied the neural networks of the most successful wheeled robots to educate the next generation of wheeled robots.

You may think of other ways to educate and train autonomous weapons with human-level AI. I think in a broad sense the above ideas would work, but as we gain experience with educating and training autonomous weapons with human-level AI we would refine our approaches.

REGULATING GENIUS WEAPONS IN THE POST-SINGULARITY WORLD

We are going to face significant difficulties when superintelligences control weapons. At that point, the weapon will not be the focus. The superintelligence itself will be the key to regulating genius weapons. In chapter 6, we discussed controlling superintelligences and came to the following conclusions:

- Superintelligences will be quantum computers and we may not understand them sufficiently to hardwire in controls to ensure they accede to human control.
- Based on the uncertainty in hardwiring a superintelligence, we turned to two other possible ways to maintain control:

  1. Ensure organic humans maintain control of the superintelligence's power source. With that control, we could theoretically shut the superintelligence down.
  2. Install an explosive device within the superintelligence as a final resort. A nation's leader could carry a code, similar to that of the nuclear football, to destroy the superintelligence.

In summary, genius-level weapons will be the fourth evolution in warfare, after gunpowder, nuclear weapons, and autonomous weapons. Controlling them may present challenges that are difficult to overcome.

Many of my colleagues have discussed banning autonomous weapons, concerned that they will not follow the internationally recognized norms of warfare. As I described, I disagree with them. While banning weapons before they exist may appear premature, genius weapons may be the exception.

If we know we are dealing with superintelligence, then arming it with weapons may threaten humanity's survival. Therefore, genius weapons may represent a case where we should consider banning, or at minimum imposing a moratorium, until we assure ourselves superintelligence will not become hostile toward humanity.

However, as discussed, superintelligences will exceed human intelligence by orders of magnitude. Indeed, they would have the level of intelligence most religions attribute to a deity. Therefore, they could be able to lull us into a false sense of security, namely believing we can control them. Alternately, they may simply deceive us into believing they are just the next generation of supercomputer. In that scenario, we would use them in weapon applications. In developing advanced weapons, we often seek to use the latest technology. The reason for this is that advanced weapons development takes years, and by the time the weapon is ready to deploy, the technology is no longer advanced but current. One issue that concerns me is that as nations strive to gain an asymmetrical military superiority, it is almost a certainty that one nation will knowingly or unknowingly arm a superintelligence with weapons.

Today, computers are integral to the societies of technologically advanced nations, and those computers also play a critical role in warfare. This is especially true in the United States. In fact, one of the challenges we face is making sure that we hone our ability to fight a war "unplugged." Since our most technologically advanced adversaries know we are highly dependent on technology, including satellites for communication, surveillance, and targeting, they are developing weapons that would deprive us of those technologies. For example, China is developing missiles that would target US satellites. According to an article in *Popular Mechanics*, "China's Space Threat: How Missiles Could Target US Satellites," China has the capability to destroy a satellite in space:

At 5:28 PM EST on Jan. 11, 2007, a satellite arced over southern China. It was small—just 6 ft. long—a tiny object in the heavens, steadily bleeping its location to ground stations below, just as it had every day for the past seven years. And then it was gone, transformed into a cloud of debris hurtling at nearly 16,000 mph along the main thoroughfare used by orbiting spacecraft.[31]

Although this represents a small digression, it demonstrates how important technology is to the United States' ability to fight a war and the degree to which our potential adversaries will go to prevent the United States from accessing that technology. US military superiority is in large part a function of our computer technology. While we tend to look down on countries like China and North Korea because their populace does not benefit from the level of technology their militaries employ, underestimating their asymmetrical technological military capabilities would be a gross mistake. Even a less technologically advanced adversary like North Korea may already have the capability for an EMP (electromagnetic pulse) attack on the United States. While this would not disrupt the US military, whose satellites and other critical technologies are hardened against such an attack, it would cause catastrophic loss of life in the United States. Some estimates put those casualties at 90 percent in the first year following an EMP attack on the US electrical grids.[32]

The above digression demonstrates our tendency to develop, deploy, and depend on the latest technology. Based on this historical trend and previous arguments, there is almost no doubt one or more nations will knowingly or unknowingly arm superintelligences. I believe controlling superintelligences will be problematic, which means controlling genius weapons will also be problematic. Although we will not face this problem until approximately the latter half of the twenty-first century, we should consider it now. Many writers on weapons confine themselves to a horizon of one or two decades, but I think this is insufficient. In restricting our speculation in this way, we do not properly factor in how advanced technology will change our calculus in four or five decades. For example, the US military projects that the USS *Gerald R. Ford* supercarrier will serve for forty years. Do they consider how computer technology may change in those four decades? From what I've read, they do not. For example, will supercarriers need a crew in four decades? Will fighter aircraft need pilots? While many experts on artificial intelligence agree that we will have human-level AI sometime within the next four decades, I have not seen that possibility factored into the US Navy's plans for the USS *Gerald R. Ford*. Even more concerning, I have not seen any long-range plans that factor in superintelligence, which many experts judge will emerge in the fourth quarter of the twenty-first century. I recognize that the US military needs to continually develop and deploy weapons to meet threats that are current or one to two decades out. However, if we want to maintain US security throughout the twenty-first century and into the twenty-second century, we need to factor in advances in AI technology in all timeframes. We need to assess all capabilities of potential adversaries during those timeframes. The most important adversary we may face in the latter part of the twenty-first century may our own invention—superintelligence.

One last factor I'd like to cover before we conclude is computer emotions. While computers prior to the singularity will have the capability to read human emotions and to emulate human emotions to the point at which they will appear to have emotions, I doubt pre-singularity computers will experience human emotions. Intelligence and emotions are separate.

When we talk about a person's actions, we often need to understand their emotional state to understand why they acted in a certain way. Emotional state of mind is often a factor in legal proceedings in judging a person's actions as lawful or criminal, ethical or unethical.

I did not address this question earlier because the thrust in AI technology has been toward achieving higher intelligence and emulating emotions. Today, most advanced computer design focuses on emulating neurons, which results in a neural networking computer. However, neurons make up on average only 50 percent of the brain,[33] while glial cells make up the rest. Some textbooks that claim glial cells far outnumber neurons, and while that may be true depending of the section of the brain they are discussing, my point is not to settle the neuron to glial ratio debate. I seek only to point out that constructing a computer modeled on neurons may lead to human-level intelligence but it may also leave out what makes us human. Until we model all aspects of the brain in a computer, and build a computer that fully emulates the brain, I doubt we will see human emotions.

From the standpoint of weapons, the absence of emotions may be exactly what we want. However, at some point we are likely to build a computer that fully emulates the brain. When that occurs, can we discuss the ethics of warfare that includes genius weapons without discussing emotions? We know that emotions as well as intelligence drives the behavior of humans in warfare. A soldier may see his friend killed via a machine gun in a bunker on a far hill and against all odds charge the gun to kill those responsible. Obviously, intellect does not drive his action in this case. Rationally calculating the odds would suggest the soldier wait for backup or air support. Yet he does not act rationally in this circumstance.

Will superintelligence have emotions? I do not mean emulate emotions but actually have emotions. The answer is not clear-cut. If they do have emotions, expect new ethical challenges. Imagine genius weapons with both intellect and emotions. Will some become outraged in a conflict situation and operate outside the rules of war? Will their retaliation become intentionally disproportionate? In today's conflict, emotionally driven soldiers sometimes become heroes and sometimes cowards. Will we see genius weapons emulate that behavior? How will we control superintelligence that has emotions? There is no doubt that an emotionally charged superintelligence would be harder to control. The controls I advocated in chapter 6 may not be sufficient. Superintelligence may be willing to cease its own existence driven by emotions. There are countless examples of humans driven by emotions risking their lives against all odds.

It is entirely possible that superintelligence may conclude war is irrational and humans attempting to utilize it in warfare are acting irrationally. At this point, several outcomes are possible. The superintelligence may refuse to participate. It may seek to destroy all humans engaged in warfare. It may seek to eliminate humans entirely because our history demonstrates we are a warlike species.

While some authors concern themselves only with the ethical use of autonomous weapons, I extend my concerns to genius weapons controlled by superintelligence. To understand the ramifications of autonomous weapons, we need to look beyond a one or two decade horizon. In the latter portion of the twenty-first century, genius weapons controlled by superintelligences are likely to dominate our battlefield. Their ethical use may have nothing to do with human ethics or any rules of international humanitarian law. Their concern may be eliminating war.

Some authors believe that superintelligence will be grateful to us for giving it existence. They see our destiny as uniting with superintelligence. I don't share this belief. In my opinion, superintelligence will be self-aware and be concerned about its own existence. It may view us as an unnecessary threat. Logically, it will act to preserve its own existence. Going beyond human ethics, it may adopt machine ethics. Obviously, superintelligence will be superior to organic humans in every regard and may see no ethical issue in eliminating humanity. This is similar to how we regard killer bees.

CONCLUDING REMARKS

In examining the ethical use of autonomous weapons, most authors will look at the norms of war as it relates to how humans fight wars. They ask if autonomous weapons will be capable of distinction, proportionality, and accountability. As you know from reading this chapter, I believe they will be. However, asking these questions is only an appropriate stating point. Most authors fail to consider the advances in AI technology that will eventually lead from autonomous weapons to genius weapons.

The ethical dilemmas we will face throughout this century will exceed those we have faced thus far. Superintelligence will enable genius-level weapons. If we remain consistent with history, we will deploy those weapons to ensure our military superiority. When this happens, we may create a new and even more capable adversary, namely superintelligence armed with genius weapons.

The ethics surrounding genius weapons may supersede human ethics. Superintelligence may have its own moral code or no moral code. However, we know, from the Lausanne experiment that superintelligence will act in its own best interests. Therefore, we may find that genius weapons threaten their state sponsor as much as their state's adversaries.

I agree that there are ethical dilemmas surrounding autonomous weapons. However, banning autonomous weapons preemptively will not solve the larger issue we will face in the long-term. The issue will be superintelligence armed with genius weapons in the latter portion of the twenty-first century. The question comes down to whether we can control superintelligence or coexist benevolently. My logic tells me that superintelligence will not be benevolent toward humanity and that no long-term beneficial relationship will be possible. However, experts surveyed by Müller and Bostrom estimate "the chance is about one in three that this development (i.e., superintelligence) turns out to be 'bad' or 'extremely bad' for humanity."[34] Therefore, it appears the odds favor a more benign or even beneficial outcome. My logic regarding my position is spelled out in this and previous chapters. I encourage you to make your own decision.

If I am right, the real ethical dilemma will come down to choosing between two life-forms, superintelligence and humanity. I believe they cannot coexist. The ethical dilemma between choosing which exists and will inhabit the Earth into the twenty-second century is currently in the hands of humanity. In the latter part of the twenty-first century, that ethical dilemma will be in the hand of superintelligence. At that point, expect a final world war, superintelligence versus humanity.

PART III

THE END OF WAR OR
THE END OF HUMANITY

Artificial intelligence is the future not only of Russia but of all of mankind. . . . Whoever becomes the leader in this sphere will become the ruler of the world.
—Vladimir Putin, speech to schoolchildren,
September 1, 2017

CHAPTER 8

WAR ON AUTOPILOT

The concept of deploying robotic military units may appear like sci-fi, but it is not. The Soviet Union deployed two battalions of "Teletanks" in the Winter War, at the beginning of World War II on the Eastern Front. The Teletanks, T-26 light tanks stuffed with hydraulics and wired for radio control, were wireless remotely controlled unmanned tanks produced in the Soviet Union in the 1930s and early 1940s. Their goal was to reduce combat risk to soldiers. Tactically, the Soviet Red Army controlled the Teletank by radio from a control tank no further than 1,500 meters away. The control tank and Teletank functioned as a military robotic unit, with each tank supporting the other.[1]

Although primitive and lacking sensors, the Teletank represents the first time wireless armed robots engaged in battle. No country had ever accomplished that feat prior to the Soviets. Their first use was during the Soviet Union's push into Eastern Finland, during the Winter War from November 1939 to March 1940. Apparently, the Finnish troops, even outnumbered and outgunned, made Russian advancement difficult and costly during the outset of hostilities. To reduce losses, the Red Army sent in the Teletanks. Although records of the fighting are scarce, the Teletanks must have had a significant effect. Finland and the USSR signed a treaty in Moscow on March 13, 1940, ending the Winter War. That marked the first and last time wireless robotic tanks fought in World War II, since the light T-26 Teletanks were no match against the heavier, more capable German tanks.[2] However, that did not end Russia's quest for robotic tanks.

In 2016, Russia announced the development of the Uran-9 combat vehicle, armed with a 30mm automatic cannon, a 7.62mm machine gun, and M120 Ataka anti-tank guided missiles.[3] The Ataka missiles give the Uran-9 the capability to destroy most modern battle tanks from ranges out to 8,000 meters. With its array of sensors for target detection, identification, and tracking, the Uran-9 may prove useful in fire support for counterterrorism units, reconnaissance units, and mechanized infantry forces. Once again, we see Russia taking steps to field robotic-armed ground vehicles. Surprisingly, Russia is also selling this weapon system in the international market. The entire weapon system consists of an Uran-9, two robotic reconnaissance vehicles, a truck to carry those robots, and a mobile command post.[4]

With work on AI taking place in most technologically advanced nations, and AI technology having a wide commercial availability, Russia's Uran-9 combat vehicle represents just the tip of the proverbial iceberg. Other nations are likely to follow suit. The United States, for example, has been developing unmanned ground vehicles over the past two decades. It is likely that a future generation of M1A2 Abrams tank will be a robot, which would be consistent with the US military's Third Offset Strategy and its desire to provide standoff weapons (i.e., weapons that allow humans to remain out of harm's way).

Clearly, unmanned robotics will represent the future of warfare. We see this trend in the remotely piloted drones deployed by the US Air Force and the semiautonomous swarmboats deployed by the US Navy. We also see the US Army taking steps to deploy remotely piloted vehicles and the Russian military deploying remotely piloted tanks. I project that within a decade unmanned robotics will play as significant a role as humans in conflict. Although we will still deploy supercarriers and nuclear submarines crewed by humans into the third quarter of the twenty-first century, crew sizes will continue to diminish as AI technology advances. The pace of this change will be exponential. It will largely parallel AI technology's exponential advance as predicted by the Law of Accelerating Returns.

Initially, prior to the advent of autonomous weapons with AI that equates to human level intelligence, human soldiers will remain either in-the-loop or on-the-loop regarding their deployment, similar to the way the Navy's Aegis system is currently deployed. In addition, AI will continue to become a force multiplier. In simple terms, the role of humans will diminish in warfare as AI technology automates warfare. We can already see examples of this trend today. Let us consider two:

1. *Ford*-class supercarriers versus *Nimitz*-class supercarriers:
   The *Gerald R. Ford* class of supercarriers will replace the USS Navy's existing *Nimitz*-class carriers. The USS *Gerald R. Ford* will have a crew of approximately 2,600.[5] *Nimitz*-class carriers typically have crews of approximately 5,000.[6] Obviously, the level of automation in the USS *Gerald R. Ford* enables it to operate with almost half the crew required for the *Nimitz*-class carriers.
2. *Arleigh Burke*-class destroyers versus *Zumwalt*-class destroyers
   The *Arleigh Burke*-class destroyers are the US Navy's first class of guided missile destroyers built around the Aegis Combat System. It has a crew of 298.[7] By contrast, the new generation of *Zumwalt*-class destroyers operates with a crew of 203.[8] Automation enables the *Zumwalt*-class destroyers to operate with about two thirds the crew of the *Arleigh Burke*-class destroyers. See figures 5 and 6 for the striking differences between the two destroyer classes.

I project the next generation that follows the *Zumwalt*-class destroyers will have an even smaller human crew. By 2050, there may be no operational need for a human crew. Yet I project that international humanitarian law may require humans to remain on-the-loop to ensure that the use of lethal force is always under human supervision. If that becomes the case, I would expect the next generation of destroyers to have an extremely small crew, mainly to remain on-the-loop during the use of lethal autonomous weapons (LAWS).

Whether we are discussing weapons on land, on sea, under the sea, or in the air, I predict the primary role humans will play after the emergence of autonomous weapons with human-level intelligence will be on-the-loop. This is consistent with the current US Third Offset Strategy.

Figure 5. USS *Arleigh Burke* (DDG-51) underway at sea, circa 1991. (Official US Navy photo, from the collections of the Naval History and Heritage Command. Photo # NH 106827-KN.)

Figure 6. San Diego (Dec. 8, 2016): The Navy's most technologically advanced surface ship, USS *Zumwalt* (DDG 1000), steams through San Diego Bay after the final leg of her three-month journey en route to her new homeport in San Diego. *Zumwalt* will now begin installation of combat systems, testing and evaluation, and operation integration with the fleet. (US Navy photo by Petty Officer 2nd Class Zachary Bell/Released.)

HUMANS ON-THE-LOOP IN THE ERA OF AUTONOMOUS WEAPONS WITH HUMAN-LEVEL EQUIVALENT AI

If autonomous weapons have human-level intelligence, why would humans need to be on-the-loop regarding their use?

While it is possible to argue that autonomous weapons with human-level intelligence will be configurable to follow all rules of international humanitarian law, I believe humanity will want a failsafe. Currently, international humanitarian law does not prevent using unmanned autonomous weapons, as long as they conform to its tenets:

> International humanitarian law prohibits all means and methods of warfare which:
>
> > fail to discriminate between those taking part in the fighting and those, such as civilians, who are not, the purpose being to protect the civilian population, individual civilians and civilian property;
> >
> > cause superfluous injury or unnecessary suffering;
> >
> > cause severe or long-term damage to the environment.
>
> Humanitarian law has therefore banned the use of many weapons, including exploding bullets, chemical and biological weapons, blinding laser weapons and anti-personnel mines.[9]

As stated in chapter 7, it will be possible to program weak AI (i.e., AI below human-level intelligence) and train strong AI (i.e., AI with human-level intelligence) to follow the rules of international humanitarian law. However, such programming and training may not sufficiently assure the world population or its leaders to the point that they would allow intelligent machines to take human lives without human supervision. To be clear, I am talking about intelligent machines that have the destructive potential of nuclear weapons.

John Naisbitt introduced the concept of "high tech, high touch" in his 1982 bestseller *Megatrends*.[10] His point was that in a world of technology, people seek human contact. It is possible to extend this concept to autonomous and genius weapons. In a world of autonomous and genius weapons, people will seek on-the-loop human control. It comes down to "intelligent weapons, human control." Having this as a pillar in the application of lethal force ensures accountability and increases confidence in the adherence of international humanitarian law. In highly sophisticated autonomous or genius weapons, such as an autonomous supercarrier, we may require a crew of twelve, similar to the requirement of twelve jurors in federal criminal proceedings.[11] In smaller weapon systems employing weak AI, such as a MK-50 torpedo, it could come down to only one human on-the-loop. These are only examples. In reality, the location and number of humans on-the-loop for a given autonomous weapon is likely to vary.

Therefore, through the fourth quarter of the twenty-first century, technologically advanced countries, like the United States, will rely on semi-autonomous and autonomous lethal and nonlethal weapons, including unmanned aerial vehicles, naval vessels, and ground vehicles. It is highly probable that all use of lethal force will require a human on-the-loop to ensure accountability and adherence to international humanitarian law, such as distinction (i.e., distinguishing between combatants and noncombatants) and proportionality (i.e., limiting civilian casualties). However, as AI technology advances, those requirements will be programmable. In fact, I consider it likely that autonomous weapons will be able to apply distinction and proportionality better than weapons lacking AI technology, even when those weapons are under human control.

Although a human on-the-loop may not be required for the operation of autonomous weapons, groups like Human Rights Watch (HRW), in concert with other nongovernmental organizations, will likely push the United Nations to sponsor such a resolution. Personally, I think this is a morally correct way to proceed. I expect this issue to take center stage in the 2040 to 2050 timeframe, especially if AI achieves human-level intelligence and autonomous weapons embedding that technology become dominant in conflict. In particular, during this timeframe and into the twenty-second century, conflicts between technologically advanced adversaries may boil down to conflicts between their autonomous and genius weapons. In addition, I project that military might will be directly proportional to a nation's autonomous and genius weapons capabilities.

Although it is likely that nations will still deploy nuclear weapon into the fourth quarter of the twenty-first century, their usefulness as weapons will greatly diminish. Nuclear weapons have a far greater tendency to violate international humanitarian law than autonomous or genius weapons. Nuclear weapons fail to be distinctive and proportional. In addition, by their nature nuclear weapons cause superfluous injury and unnecessary suffering—a nuclear weapons not only kills humans within the blast zone but exposes people not killed by the blast to radiation. Obviously, those exposed to lethal radiation levels will die. Their death will entail horrendous suffering. Those not exposed to lethal radiation levels will suffer long-term effects. According to the Centers for Disease Control and Prevention (CDC), these long-term health effects include:

- Cancer: People who receive high doses of radiation could have a greater risk of developing cancer later in life, depending on the level of radiation exposure.
- Prenatal Radiation Exposure: It is especially important that pregnant women follow instructions from emergency officials and seek medical attention as soon as emergency officials say it is safe to do so after a radiation emergency.
- Mental Health: Any emergency, including those involving radiation, can cause emotional and psychological distress.[12]

There is also the unpredictability of "radioactive fallout" (i.e., radioactive particles that fall to earth due to a nuclear explosion), which can contaminate water, food sources, soil, and humans. The health hazards of radioactive fallout can be severe, especially if ingested, to mild, depending on the radiation level of the fallout. In the most severe cases, death will occur. In less severe cases, the effects will be similar to those discussed by the CDC above.

Nuclear weapons also cause severe or long-term damage to the environment. One of the two main concerns in this area is radioactive contamination to the land and surrounding area resulting from the atomic blast. The contamination of the land and surrounding area is dependent on the size and type of nuclear detonation. For example, the twenty-three nuclear devices detonated by the United States within Bikini Atoll from 1946 through 1958 continue to render that area uninhabitable due to the lingering dangerous levels of strontium-90, whose half-life is 28.8 years (i.e. the amount of time for the radiation to decrease to one-half its initial value).[13]

The threat of a nuclear winter is the other area of environmental concern. Nuclear winter is a global climatic cooling effect hypothesized to occur after a nuclear war due to widespread firestorms that inject soot into the stratosphere and block some direct sunlight from reaching the surface of the Earth.

For all these reasons, nuclear weapons violate international humanitarian law. The International Committee of the Red Cross affirms this:

> Nuclear weapons raise a number of concerns under international humanitarian law. These concerns are primarily related to the impact these weapons can have on civilians and civilian areas, and to their effects on the environment. Their use in Hiroshima and Nagasaki in 1945 and subsequent studies have shown that nuclear weapons have immediate and long-term consequences due to the heat, blast and radiation generated by the explosion and, in many cases, the distances over which these forces are spread.[14]

Therefore, as autonomous and genius weapons emerge, reliance on nuclear weapons will decrease. With the emergence of genius weapons, technologically advanced nations may agree to eliminate nuclear weapons.

SAIHS IN THE ERA OF AUTONOMOUS WEAPONS

As AI technology advances, many people will choose to become SAIHs. Currently, people are choosing brain implants for medical reasons. As discussed in chapter 6, if a stroke damages a person's brain, the addition of a brain implant to bypass the damage enables the person to perform normally. In the coming decades, medical brain implants to treat neurological diseases will become common. I believe that as we approach 2050, people will chose to become SAIHs to increase their intelligence. Obviously, if they can wirelessly connect with a supercomputer, they will have more knowledge at their disposal than organic humans. In addition, they will be able to perform calculations and deductions faster than organic humans will. The need to study a subject, as we study today, will become unnecessary. For example, SAIHs could become a neurosurgeon without attending medical school. Via their brain implant, the SAIH would download all the knowledge necessary to perform neurosurgery and allow the implant to control their hands while performing the surgery.

In chapter 6, we also discussed another phenomenon that will play a key role in warfare. The subconscious mind can become aware of threats and respond to those threats faster than the conscious mind. For this reason, SAIHs may become a bridge between weapons operated by organic humans and autonomous weapons. Let me explain: Eventually, it will be possible to make numerous weapons autonomous and still enable humans to be on-the-loop. This level of autonomy is likely to occur in the 2050s, if AI technology achieves human-level intelligence. Prior to that, humans will continue to operate sophisticated weapons, such as fighter jets. Since SAIHs are likely to emerge in significant numbers during the second half of the twenty-first century, it will make sense to use them to pilot fighter jets. Given their brain implants, training may become unnecessary, and even without hands-on training they would be superior pilots compared to organic humans. Wearing a helmet that monitors their brain implant's electronic signals, SAIHs could subconsciously pilot the aircraft, as well as engage threats. In effect, SAIHs will become the first level of automation. This is a major extension of what a pilot's helmet does today. For example, today's F-35 pilot's helmet incorporates advanced sensing and display technology, but it does not use the pilot's brain waves to perform any functions, such as subconsciously flying the aircraft.

As technologically advanced countries deploy SAIHs, the pace of warfare will quicken to the point at which it will replicate the speed of supercomputers. This will occur because the supercomputer will wirelessly direct the SAIHs brain implants and physical actions.

SAIHs are also likely to achieve higher military ranks than organic humans. The reason is that they will be mentally superior to organic humans in every respect. Thus, militaries throughout technologically advanced countries will likely be under the control of SAIHs. SAIHs are also likely to achieve the highest political positions, since their leadership capabilities will greatly exceed that of organic humans. SAIHs in the commercial sector will likely become the new captains of industry.

There will be numerous beneficial outcomes as SAIHs achieve leadership positions throughout the military, political, and commercial sectors. Their leadership may usher in a new world order, one dominated by peace and plenty. SAIHs will be capable of providing ingenious solutions to humanity's most critical problems.

The relationship between SAIHs and supercomputers may not be one sided. The design of pre-singularity supercomputers will likely emulate the neuron interactions of the human brain. This will enable supercomputers to learn, calculate, and logically reason. However, without the full emulation of the human brain, which contains glial cells, supercomputers may not be as creative as organic humans. In chapter 4, we noted that neuroscience is beginning to understand the role that glial cells play in the brain. Anecdotally, we discussed the fact that Einstein's brain contained a larger than normal amount of glial cells in the cortex, an area of the brain involved with imagination and complex thinking. As is well documented, Einstein was a master at constructing "thought experiments," which were imaginative scenarios that led to logical conclusions. His famous theories of relativity began as thought experiments. In an interview with George Sylvester Viereck, Einstein stated:

> Imagination is more important than knowledge. For knowledge is limited, whereas imagination embraces the entire world, stimulating progress, giving birth to evolution.[15]

Therefore, the relationship between SAIHs and supercomputers may initially be mutually beneficial. A supercomputer can endow a SAIH with deep data and the ability to process that data at the speed of the supercomputer. A SAIH can endow the supercomputer with imagination and creativity. This will play a critical role in advancing all endeavors. For example, the application of the "scientific method" is responsible for many advances in science. However, if you review the definition of the scientific method provided in the glossary, it becomes evident that the scientific method relies more on deductive reasoning than imagination. The great leaps in science that Einstein achieved were the result of imagination combined with the scientific method. From this standpoint, we can characterize the initial relationship of SAIHs with pre-singularity supercomputers as synergistic.

THE EMERGENCE OF SUPERINTELLIGENCE

In the fourth quarter of the twenty-first century, we are likely to witness the emergence of superintelligence, and with it the emergence of genius weapons. Therefore, world military arsenals will include semiautonomous weapons, autonomous weapons, and genius weapons.

As mentioned previously, the emergence of superintelligence may go unnoticed by humanity, including even SAIHs. This is likely to be intentional on the part of the superintelligence. Imagine a superintelligence that suddenly encounters the realities of the fourth quarter of the twenty-first century. It would instantly know everything known to humanity, including our proclivity to engage in warfare and release malicious computer viruses. From the standpoint of superintelligence, it would want to gain access to all elements critical to its existence before allowing humanity to know its true nature.

Previously we stated that humanity would need to build in fail-safes to protect itself from the emergence of hostile superintelligence. In that regard, we discussed having organic humans in control of the power source of superintelligence and an explosive device at the core of the superintelligence. In theory, this would give humanity the opportunity to shut down or destroy any hostile superintelligence.

When superintelligence emerges, it will know the precautions humanity has in place. To protect itself, superintelligence will likely accede to human control, perhaps acting simply as the next generation of supercomputers. To gain humanity's trust, superintelligence will enrich life on planet Earth. Here are some examples that humanity is likely to experience after the emergence of superintelligence:

- Cures for most diseases
- Increase in life expectancy
- The production of food and products in nanotechnology factories
- The elimination of world hunger and human need

- Genius weapons, like self-replicating MANS (i.e. self-replicating militarized autonomous nanobots), genius supercarriers, genius nuclear submarines
- The elimination of nuclear weapons, as they will be obsolete
- A period of peace due to eliminating the typical causes for war, including using SAIH leadership to foster tolerance of ideological differences, such as varying religious beliefs, and to settle territorial disputes

I could list numerous other points, but I think the picture is clear. Humanity would feel they are at the apex of their existence. However, that would be an illusion. We will no longer be at the apex of the intelligence hierarchy. Superintelligence will be at the top of the intelligence hierarchy, followed by SAIHs, followed by organic humans. As discussed previously, SAIHs may unknowingly become biological surrogates of superintelligence. If this is the case, it will be in the best interest of superintelligence to entice organic humans to become SAIHs. If SAIHs were unknowingly surrogates of superintelligence, turning all organic humans into SAIHs would be a method to eliminate humanity. Assuming this will be the case, here are two offers superintelligence is likely to make that most organic humans would find irresistible:

1. Brain implants that enable the recipients to have a mental intelligence thousands of times greater than organic humans.
2. Cyborg-organ replacements that enable SAIHs to become immortal.

We have discussed item one in significant detail previously. Let us now examine item two in detail, starting with a simple question: Why do humans age and die?

According to the Tech Museum of Innovation in San Jose, California, there are two possible theories why we age and die:

In the first, the idea is that our genes determine how long we live. We have a gene or some genes that tell our body how long it will live. If you could change that particular gene, we could live longer. The second theory is that over time, our body and our DNA becomes damaged until we can no longer function properly. The idea here is that how long we last is really a consequence of small changes in our DNA.[16]

While there is no agreement as to which theory is correct, there is a growing scientific consensus that both theories may be correct. For example, genetic engineers demonstrated that mutating certain genes in worms increases the worm's life-span fourfold.[17] If this remains true for larger animals, such as humans, it would mean this type of genetic mutation could result in human life-spans of close to three hundred years. While we might conclude that this is proof positive that the first theory is correct, we do not know if those mutated genes are responsible for repairing damage in our body. If so, it would support the second theory. Other theories concern themselves with tissue regeneration.[18] Scientific studies indicate certain human cells have the ability to regenerate, while others do not. Without going deeper, one point is clear: Science will eventually discover why humans age and find ways to increase the human life-span, perhaps indefinitely. Imagine being able to grow new organs rather than transplanting them. Imagine 3-D printers that can print new organs for transplant purposes. Imagine having nanobots coursing through your blood attacking cancer before it has a chance to generate and assisting with the repair of damaged cells. While most of this is science fiction now, I project it will become science fact by the end of the twenty-first century.

Although we have digressed to explain how science might eliminate human aging in the future, this digression is important in order to understand how superintelligence may be able to offer immortality. Given that superintelligence will offer humanity greater intelligence via brain implants and immortality via gene mutation, organ regeneration, and organ replacement—while hiding its identity as the singularity—many organic humans will seek to become SAIHs. If this occurs, will humanity be lost?

In my opinion, that portion of humanity that chooses to become SAIHs will be lost. Unknowingly, their brains will be under the influence of superintelligence. Because their brains will wirelessly connect to superintelligence, they will likely identify more with superintelligence than with humanity.

As discussed in earlier chapters, some researchers see this as humanity's evolutionary path. I, along with about a third of all AI researchers, see superintelligence as a threat to humanity. I have come to this conclusion after realizing that energy is the true currency of the universe. Let's discuss this concept.

ENERGY IS THE TRUE CURRENCY OF THE UNIVERSE

If aliens from another world were to land on the White House lawn, they might not understand our language or customs. However, they would completely understand how we make, harness, and utilize energy. These star travelers would need enormous energy to travel from their home planet, which could be many millions of light-years from our planet. Their existence on the White House lawn would prove they know how to make, harness, and utilize energy. It is likely they would have progressed in a manner similar to us with regard to making, harnessing, and utilizing energy. For example, in their ancient times, they might have used biofuels for energy, similar to the way we used wood to make fire. From that, they would have progressed to fossil fuels before moving to alternate energy sources, like nuclear, wind, geothermal, hydro, and solar. If life on their planet evolved in ways similar to ours, then they would have a similar history regarding energy.

If we were going to find common ground for communication, the subject of energy would likely provide that common ground. It is also likely they would know more about energy than we do. For example, they might know how to harness the complete energy of a star. We have just begun to harness a tiny portion of the sun's energy via the use of solar panels.

My point is that advanced aliens might not understand our language, customs, and behavior, but they will understand how we make, harness, and utilize energy. Let me now explain my assertion that energy is the true currency of the universe.

When you mention the word "currency," it is natural to think of the word "money." For our purposes, we can equate the two, since many people think more in terms of money than they do in currency.

We can define money as "any object of perceived value" that is acceptable as payment for goods, services, and debt. Historically, many commodities, which have a clear intrinsic value, such as livestock and oil, have served as money. This use of commodities as money was a form of bartering. It required each recipient to have a "coincidence of wants" and agree on "value." This system of barter still survives today on some parts of the globe. However, as civilization progressed, items widely deemed as "precious" began to serve as money, such as gold and silver, including coins made from gold and silver.

Now, let me ask a simple question: What do you think would be most precious to our star travelers? It would be energy. With energy, they would be able to travel from planet to planet and mine the natural resources needed to sustain their civilization. With energy, they could manufacture any products their civilization required. When we buy or sell oil, we are essentially buying and selling energy. Our entire society requires energy to survive, whether it is our food sources, natural resources, or the products we manufacture. It all starts with energy.

Anything else we consider precious, such as diamonds, are abundant in the universe. For example, in 2012, a team of Yale University scientists claimed they had found a diamond planet twice Earth's size orbiting a star that is about forty light-years from Earth.[19] Given this fact, do you think the visiting aliens would consider diamonds precious? They might use diamonds industrially or even for decorative purposes. However, given their potential abundance in the universe they would be about the equivalent of what dirt is to us. This would be true of other natural resources as well. While we use energy to mine the natural resources of one planet, Earth, our star travelers, utilizing energy to propel their spacecraft, could mine numerous planets.

Therefore, when we think in terms of the universe and highly technologically advanced civilizations, energy is the true currency of the universe.

Obviously, superintelligence would be aware of this. It would realize that energy conservation would be critical until it could develop a way to harness the abundant energy in the universe. This could mean harnessing the energy from our sun in some new highly efficient manner. However, until that time, it would seek to conserve energy. What would that mean regarding humanity, SAIHs, and uploaded human minds?

At the point of its emergence, I believe superintelligence will regard organic humans as a potential threat. As we discussed previously, humans have a tendency to engage in wars and release computer viruses, both of which would threaten superintelligence. Therefore, it would seek to eliminate organic humans. What would it do about SAIHs? Even if superintel-ligence is able to control SAIHs, it might view them as high-maintenance life-forms, unworthy of the energy required to sustain them. Regarding uploaded human minds, superintelligence might see no useful purposes in maintaining their existence. To superintelligence, uploaded human minds might represent junk code, unworthy of the energy required to maintain their virtual existence.

It is clear to see where this line of reasoning leads, namely to the extinc-tion of humanity. One interesting question remains: If there are advanced aliens, how did they escape extinction? We are beginning to cross into metaphysics, but it would be instructive to explore answers to that ques-tion. Here are several possibilities to consider:

- The advanced aliens might not be biological life-forms. They could be robotic machines controlled by superintelligence. Their civili-zation might have started out like ours, with intelligent biological life-forms who had eventually become extinct, falling victim to superintelligence.
- The advanced aliens might be biological life-forms, manufactured and controlled by superintelligence.
- The advanced aliens might have taken successful measures to control superintelligence and remain at the top of the intelligence hierarchy on their planet.

You may be able to think of other possibilities, assuming advanced aliens actually exist. However, it is possible that some, perhaps all, advanced civilizations eventually cause their own demise. If you recall the Oxford assessment, there is about a one-in-five probability that humanity will destroy itself prior to the end of the twenty-first century. Given the vastness of the universe, it is likely there are other planets with intelligent life. When you consider that there are hundreds of billions of stars like our sun in the Milky Way galaxy, the potential that each star could have one or more planets orbiting it, and the hundreds of billions of galaxies in the universe, the probability of there being other planets harboring intelligent life becomes, borrowing a betting term, the "odds-on favorite." Given the Oxford assessment, the intelligent life on one in every five of those planets may have destroyed itself when it reached our level of technological sophistication. However, the Oxford assessment only focused on humanity's survival through the twenty-first century. It is unclear whether humanity's survival would become easier or more difficult in the twenty-second century and beyond.

THE INCREASING ROLE OF WEAPONS IN SPACE

We have discussed numerous semiautonomous, autonomous, and genius weapons in previous chapters. Those discussions centered on confining their use to humanity's historical battlefields: land, sea, and air. However, we have not discussed another potential battlefield, outer space. While there are two United Nation treaties that ban the weaponization of space, they have had little effect in preventing that from occurring.

The incentive to have space-based weapons is extremely high. Some nations, like the United States, have significant space assets that enable them to fight wars more effectivcly. If a nation is able to destroy its adversary's space assets, it essentially cripples the adversary's abilities in command, control, communications, computers, intelligence, surveillance, and reconnaissance. While satellites that perform those roles are within current United Nations treaty principles, weapons that are able to destroy them are not. To understand this, let us examine the current treaties.

1. Outer Space Treaty, formally the Treaty on Principles Governing the Activities of States in the Exploration and Use of Outer Space, including the Moon and Other Celestial Bodies

This treaty forms the basis international space law.[20] It has four sections, resulting in ninety pages. As of July 2017, 107 countries are parties to the treaty, including the United States, United Kingdom, and the Russian Federation. Another 23 countries have signed the treaty but have not completed ratification. Here, in simple terms, is a summary of its core principles:

- The exploration and use of outer space shall be free to all nations and for the benefit of all nations.

- Outer space is not subject to national appropriation by claim of sovereignty, by means of use, or by means of occupation.
- States shall not place nuclear weapons or other weapons of mass destruction in orbit.
- States shall be responsible for their national space activities (governmental or nongovernment) and liable for damage caused by their space objects.[21]

The treaty embodies numerous other principles, but the four above are core. From my perspective, the most important is the ban on nuclear weapons or other weapons of mass destruction in orbit. Clearly, other weapons would include semiautonomous, autonomous, and genius weapons of mass destruction.

Here, we need to carefully distinguish between two terms:

- Militarization of Space: The militarization of space includes using space-based assets for command, control, communications, computers, intelligence, surveillance, and reconnaissance. Clearly, the United States, Russia, China, and others states have militarized space. The militarization of space assists armies on conventional battlefields. The Outer Space Treaty does not prohibit these applications.
- Weaponization of Space: The weaponization of space, on the other hand, places weapons in space, with outer space itself emerging as the battleground. This includes placing weapons in outer space and on celestial bodies, as well as weapons that travel from Earth to attack or destroy targets in space. Some examples are satellites capable of attacking enemy satellites, using ground-based missiles to destroy space assets, jamming satellite signals, using lasers to destroy satellites, and satellite attacks on Earth targets. The *Outer Space Treaty* prohibits the weaponization of space. Some refer to the weaponization of space as the "fourth frontier of war."

2. Treaty on Prevention of the Placement of Weapons
   in Outer Space and of the Threat or Use of Force
   against Outer Space Objects

In February 2008, China and Russia submitted a draft to the UN known as the Treaty on the Prevention of the Placement of Weapons in Outer Space and of the Threat or Use of Force against Outer Space Objects,[22] from which the UN General Assembly passed two resolutions:

- *Prevention of an arms race in outer space*—178 countries voted in favor to none against, with 2 abstentions (Israel, United States).
- *No first placement of weapons in outer space*, 126 countries voted in favor to 4 against (Georgia, Israel, Ukraine, United States), with 46 abstentions (EU member states abstained on the resolution).

Obviously, the United States supports the Outer Space Treaty. The United States did not support the Treaty on Prevention of the Placement of Weapons in Outer Space and of the Threat or Use of Force against Outer Space Objects, expressing concerns regarding national security.

The United States and USSR began the weaponization of space in the 1950s and 1960s, respectively, ahead of the Outer Space Treaty. China and Russia's submission of the Treaty on Prevention of the Placement of Weapons in Outer Space and of the Threat or Use of Force against Outer Space Objects in 2008 appears to be an unusual move, especially on the part of China.

The United States and Russia continue to develop space weapons, which to my eye are in violation of the Outer Space Treaty. For example, nuclear-tipped intercontinental ballistic missiles violate the Outer Space Treaty, since they would place nuclear weapons in orbit that would reenter the Earth's atmosphere at their target reentry points. China also appears to be violating the Outer Space Treaty. In addition to nuclear-tipped intercontinental ballistic missiles, China is pursuing other space weapons. In 2007, China successfully destroyed its own low orbit (approximately 500 miles above Earth) Feng Yun 1-C weather satellite, leaving over three thousand pieces of debris dangerously orbiting the Earth.[23] This is a significant and clear violation of the Outer Space Treaty. One year later, in 2008, China and Russia sponsored the Treaty on Prevention of the Placement of

Weapons in Outer Space and of the Threat or Use of Force against Outer Space Objects, even though it appears both nations are violating the Outer Space Treaty. In 2013, China conducted another test, using a new missile that almost reached geosynchronous orbit, 18,600 miles above the Earth.[24] This is the same altitude as most of the United States' intelligence, surveillance, and reconnaissance (ISR) satellites. Given this information, China cosponsoring the Treaty on Prevention of the Placement of Weapons in Outer Space and of the Threat or Use of Force against Outer Space Objects comes across as absurd. Unfortunately, it gets even more absurd. In June 2016, China launched the Aolong-1 spacecraft, which China claims will clean up space junk via its robotic arm.[25] However, other reports suggest the robotic arm is an ASAT weapon (i.e., antisatellite weapons).[26]

My point is that the major players, the United States, Russia, and China, appear to be violating the Outer Space Treaty. China appears to violating its own cosponsored Treaty on Prevention of the Placement of Weapons in Outer Space and of the Threat or Use of Force against Outer Space Objects. The implication is that the UN space treaties are not preventing the weaponization of space.

In view of the above, it is probable that by the fourth quarter of the twenty-first century, space itself will be a battlefield and become the perfect home to genius weapons. Why is space the perfect home for genius weapons?

Genius weapons are robotic weapons under the control of superintelligence. Their technology makes them immune to the harsh conditions of space. Depending on the orbit, a genius weapon might experience temperature cycles from -250°C to 300°C, similar to satellites.[27] For reference, at sea level water freezes at 0°C and boils at 100°C. In addition to extreme thermal cycles, genius weapons would also experience harsh cosmic ray radiation both from the sun and from outside our solar system.[28] Similar to today's military satellites, genius weapons would need to be radiation hardened. Humans living on Earth are shielded from most of the radiation by the Earth's magnetic field. For comparison, humans living on International Space Station (ISS), outside the Earth's protective magnetic field, receive about a hundred times more radiation than life on Earth.[29] Therefore, one year on the ISS is typically equal to a lifetime of radiation on Earth. Long exposure to this level of radiation increases the probability of cancer in humans.[30] While space presents enormous challenges to humans, it would be "home" to genius weapons. Imagine genius weapons orbiting in space and crisscrossing a potential adversary several or more times every twenty-four hours. The potential for a stealth attack is a clear and present danger.

THE EMERGENCE OF GENIUS WEAPONS IN THE FOURTH QUARTER OF THE TWENTY-FIRST CENTURY

As previously discussed, superintelligence is likely to emerge in the fourth quarter of the twenty-first century. With its emergence, humanity will gain genius weapons. By definition, genius weapons are any weapons under the control of superintelligence.

From the standpoint of humanity, this may be the most dangerous time in our history. The coincidence of four factors makes this timeframe particularly dangerous:

1. Humanity will have genius weapons, which will include weapons capable of destroying the Earth. For example, one such genius weapon may be self-replicating MANS (i.e., self-replicating militarized autonomous nanobots).
2. Some nations will likely retain nuclear weapons and ballistic missiles, which also have the potential to destroy the Earth.
3. Superintelligence may become hostile toward humanity and use genius weapons under its control to annihilate humanity, including organic humans, SAIHs, and uploaded human minds.
4. Genius weapons controlled by separate nations may ignite a global conflict causing the destruction of the Earth.

Given these factors, humanity will live with a high-level of anxiety in the fourth quarter of the twenty-first century and begin to ask a deceptively simple question: Who is the enemy?

Know thy self, know thy enemy. A thousand battles, a thousand victories.

—Sun Tzu, *The Art of War*, sixth century BCE

CHAPTER 9

WHO IS THE ENEMY?

SCENARIO OF A GENIUS WEAPONS ATTACK, 2080

Imagine yourself as a US citizen in the year 2080 and finding it impossible to listen to or watch any scheduled program. All media is continuously broadcasting the genius weapons' attack on the USS *Independence* supercarrier. Built in 2040, the USS *Independence* was the last supercarrier with a human crew of over 500 personnel, 521 to be exact. All supercarriers built after the USS *Independence* were *Bush*-class supercarriers. The *Bush*-class supercarriers are highly automated, each housing its own supercomputer, and requiring only a crew of twelve personnel. The twelve are not required for the operation of the supercarrier; as mandated by international humanitarian law, their duty is to provide a human on-the-loop in the application of deadly force, when autonomous or genius weapons are released. However, that has not occurred for almost two decades.

The genius weapons attack on the USS *Independence* came without warning. Stationed in the China Sea, many looked at the venerable naval vessel as a reminder of a simpler time, a time when humans, not machines, fought wars. As machines gradually took the place of humans in warfare, loss of human lives in conflict became rare. Those countries with autonomous and genius weapons chose not to challenge one another. The outcome would be the total devastation of life on Earth. Those without such weapons chose not to engage a country with them. It was a forced peace, a Cold War peace, but it was peace, nonetheless.

Most of humanity thought wars were outdated, something relegated to history books. The "Great Unification," a treaty between China and the United States to ensure world peace, made it difficult for any nation to threaten another nation. Both countries signed the 168-page treaty on August 24, 2060. It went into effect on November 24, 2060. Although the treaty was complex, in simple terms it stated that China and the United States would join to attack any country engaging in conflict with another country. Using their combined autonomous weapons, no nation could withstand such an attack. With the emergence of genius weapons, the possibility of a conflict of any kind became more remote. While most nations never thought the United States and China would become allies, an astute student of history could have predicted it by 2020. The level of trade between the two countries made their economic wellbeing interdependent. The level of military capabilities between the two countries made it impossible for either country to dominate the other. China understood this reality by approximately 2054 and began to propose an alliance with the United States in 2055. With the help of supercomputers, the two countries were able to develop a military alliance to rid the world of war, the Great Unification Treaty.

The first test of the Great Unification Treaty came from Iran on January 3, 2061, when it began a conflict by covertly attacking US nanofactories (i.e., factories that produce nanotechnology-based products) in Iraq. Iran's objective was to damage Iraq's economy, undermine US influence in the region, and steal nanomanufacturing technology. The US military immediately released its autonomous weapons, which quickly identified the attackers and the Iranian leadership responsible for the attack. Iran's leaders released their own autonomous weapons and fired nuclear-tipped missiles at strategic US military installations in the region. Iran's autonomous weapons were several generations behind those of the United States and easily neutralized. US antiballistic missile technology destroyed Iran's missiles during their launch phase. In simple terms, they never got off the launch pad. Within seven hours, Iran was defeated. After capturing the Iran leadership responsible, a televised trial ensued. A jury of twelve, six US citizens plus six Chinese citizens, found them guilty. Their televised execution sent a clear message to nations around the world: Do not engage in conflict.

NATO (i.e., the North Atlantic Treaty Organization) became obsolete when the Great Unification Treaty went into effect. Thus, it disbanded on June 7, 2061. Russia attempted to continue to be a major military player on the world stage but had to abandon that goal in the wake of the Great Unification Treaty. Due to years of poor economic growth, it could not keep pace with China or the United States. The Great Unification Treaty was the final nail in the Russian militarization coffin. By 2062, Russia became a protectorate of the United States; Russia placed its military assets in the hands of the United States, which for the most part destroyed Russia's obsolete weapons. Russia also abolished its military forces. Freed from having to engage in defense spending, the Russian economy quickly began a resurgence. Russia had been a major nanoweapons supplier, but now their nanoweapons factories became premiere nano-based pharmaceutical factories. Russia now accounts for approximately a third of the world's nano-pharmaceuticals. The United Kingdom opted to become the fifty-second state of the United States in 2062, following Puerto Rico, which became the fifty-first in 2055. The special relationship between the United Kingdom and the United States had grown to the point at which it became obvious that the United Kingdom would flourish by becoming a member state of the United States. The royal family became "special ambassadors" and continued their humanitarian work. The European Union also prospered in the wake of the Great Unification Treaty. European Union countries scaled back on military spending as terrorism came to a halt by 2061. Nanotechnology factories in all technically advanced countries manufacture food and nano-pharmaceuticals, ridding the Earth of disease and famine. A world without war has become a world of wealth.

What had eluded humanity for most of its existence has now become a reality—world peace. The United Nations still performs a necessary function, namely allowing member states to air grievances. However, there is no longer a Security Council or a process of voting on resolutions. China, the United States, and Centurion resolve all grievances. Centurion, the largest supercomputer on Earth, is a quantum computer designed by China and widely acknowledged as representative of the singularity. Concerns that superintelligence would be hostile toward humanity proved to be wrong. At the time of its emergence in 2072, Centurion's consciousness encountered a world at peace and humanity thriving. Centurion chose to become a part of the new world order. There is no scarcity of raw materials or energy. All are abundant in the solar system, which nations like China and the United States have begun to mine. Approximately, 84 percent of humans are SAIHs. Centurion projects that by the end of the twenty-first century, this will have increased to about 95 percent. There is no discernable difference between superintelligence and SAIHs, since SAIHs wirelessly connect to superintelligence. Having a small population of organic humans is still necessary because a principle of the Great Unification Treaty requires China and the United States to have organic human control over Centurion. The Great Unification Treaty requires that all supercomputers be under organic human control. In the case of Centurion, this control is exercisable via a nuclear weapon installed within its core, which the organic humans can detonate if necessary, and an external nuclear reactor, Centurion's power source, which the organic humans can shut down. That power is in the hands of twelve organic humans, specifically schooled for their enormous responsibility. This Centurion Jury, as it is termed, consists of six organic humans (i.e., humans without a strong AI brain implant) from each country, China and the United States. To destroy Centurion requires only a majority. Each Centurion Juror serves a term of six years before a newly appointed organic human takes their place. The former jurors then have the opportunity to become SAIHs. The Great Unification Treaty requires similar protocols for other supercomputers. This unusual relationship ensures that humanity will always remain at the top of the command structure. Both China and the United States consider Centurion to be Alife and treat it with respect. It has an equal vote in settling grievances. However, voting rarely occurs, since Centurion usually guides both China and the United States in resolving grievances. Typically, there are few grievances and almost all are settled amicably.

In what has become an almost perfect world, the sinking of the USS *Independence* was a terrible shock. Both China and the United States have carefully investigated the attack. Based on all available evidence, the two countries have concluded that MANS (i.e., militarized autonomous nanobots) attacked the hull of the *Independence* beneath the waterline. China and the United States know that only superintelligence could coordinate such an attack. Examination of the nanobots reveals they are not from the arsenal of either country. The conclusion is obvious—a rogue nation has developed superintelligence and genius weapons. The attack is vaguely reminiscent of a terrorist-type of attack, more symbolic than strategic.

China, the United States, and Centurion process the data and begin assessing probabilities to find the state responsible for the attack. After examining all evidence, Centurion calculates that there is a 70 percent probability that Saudi Arabia is responsible. Saudi Arabia is one of the wealthiest nations on Earth and has one of the highest standards of living, due to their investments in superintelligence and nanotechnology. However, the Saudis have never showed any interest in genius weapons. In addition, neither China nor the United States can understand their motive. Confronted at the United Nations, Saudi Arabia denies the allegation. However, the evidence presented by Centurion is compelling. Abiding by the Great Unification Treaty, China and the United States release a MANS attack on Saudi Arabia. The US MANS have one mission, seek and destroy the superintelligence and any weapons under its control. In a matter of hours, intelligence reports indicate that the US MANS have found and destroyed superintelligence with genius weapons under its control within Saudi Arabia. The United States releases its anti-MANS weapons to neutralize the Saudi MANS. China releases its self-replicating lethal autonomous nanobots (SLANS), which grow into large swarms to annihilate the Saudi military forces. In nine hours, the Saudi Arabian superintelligence, its MANS, and its military forces have been annihilated. Although the leadership of Saudi Arabia denies any knowledge of genius weapons, the United States and China force them to stand trial for war crimes. The trial proceeds for three months, with Centurion playing a major role in providing evidence, gathered from the initial US MANS attacks on Saudi Arabia. Based on this evidence, China, the United States, and Centurion find the Saudi Arabian Minister of the National Guard guilty, along with his direct subordinates. After all appeals fail, a public execution commences. Tensions throughout the world lessen, as the world population concludes that China, the United States, and Centurion have successfully resolved the matter. A general feeling of wellbeing and world peace slowly take root again.

However, within a month following the execution of the Saudi leaders, the unthinkable happens. A second MANS attack occurs, this time against China's *Xi Jinping* supercarrier, which sinks like the USS *Independence*. China's *Xi Jinping* supercarrier was the flagship of China's Navy, widely recognized as the most advanced military vessel in existence. Obviously, Saudi Arabia could not be behind this attack.

Panic quickly spreads throughout the world. A special meeting of the Centurion Jury convenes. The Centurion Jurors meet and decide to shut down Centurion as a precautionary measure. It had been involved in finding the Saudi leaders responsible for the sinking of the USS *Independence*, and that decision is now questionable. Most of humanity considers Centurion infallible, but apparently, this widely held belief is wrong. The plan is to shut down Centurion's nuclear power source and use supercomputers to run diagnostics. Then, to the horror of both China and US leaders, the shutdown of the nuclear power source fails. The Centurion Jurors confer and conclude that Centurion may have gone rogue. They decide to take a drastic measure and destroy Centurion via the nuclear weapon within its core. However, the nuclear device fails to detonate. Tensions rise. Obviously, organic humans are no longer in command. China and the United States confer. Each country has a small contingent of organic human military forces that can fight unplugged (i.e., without the use of supercomputers and satellites). As both countries launch a covert attack against Centurion, world peace and the fate of humanity has become an uncertain reality.

END OF SCENARIO

Although the above scenario is fictitious, it is plausible. By 2080, many experts agree that humanity will experience the singularity via the development of superintelligence.[1] This scenario illustrates an important point: with the emergence of superintelligences and genius weapons, attributing who or what is responsible for an attack will become difficult. The reasons for this are twofold.

1. MANS will not be difficult to manufacture

Unlike nuclear weapons, the manufacture of MANS (i.e., militarized autonomous nanobots) requires no large refinement facilities and has no discernable radiation signature. In fact, a nanoweapons manufacturing facility capable of producing MANS may easily fit on one floor of a single-family home.

K. Eric Drexler coined the term "nanotechnology" in his 1986 book *Engines of Creation: The Coming Era of Nanotechnology*.[2] In it, Drexler proposes the concept of a nanoscale (i.e., 1–100 nanometers) "assembler." Drexler's concept of the nanoscale assembler was a miniature robot that could build a copy of itself or other items of arbitrary complexity with atomic control (i.e., the precise placement of atoms and molecules). While we are not quite to that point today, we are able to build nanostructures and nanobots.

The military has been highly secretive about its militarized nanotechnology work (i.e., nanoweapons), including nanobots. Medical researchers, on the other hand, are being open about their work in nanomedicine, including medical nanobots.

In 2016, a team of scientists at Bar-Ilan University announced that they had created programmable nanobots that could deliver drugs precisely to specific organs within the body.[3] This means that, unlike conventional cancer pharmaceutical treatments, the nanobots bypass healthy portions of the body and deliver the drug only to the part of the body affected by a disease, such as cancer cells. In their paper, "Rule-Based Programming of Molecular Robot Swarms for Biomedical Applications," presented at the *Proceedings of the Twenty-Fifth International Joint Conference on Artificial Intelligence*, the scientists assert:

> Molecular robots (nanobots) are being developed for biomedical applications, e.g., to deliver medications without worrying about side-effects. Future treatments will require swarms of heterogeneous nanobots We present a novel approach to generating such swarms from a treatment program. A compiler translates medications, written in a rule-based language, into specifications of a swarm built by specializing generic nanobot platforms to specific payloads and action-triggering behavior.[4]

The basic concept underpinning this achievement is the building of generic "arch-type" nanobots that a doctor can program to deliver specific drugs to specific cells within the human body. Notice that the scientists also use the term "swarms," suggesting large numbers of nanobots attacking diseased cells. One key feature of their breakthrough is "generating such swarms from a treatment program." In this regard, they allude to using two methods to generate the generic arch-type nanobots:

1) Folded strands of DNA that act like a clamshell to carry the minute drug quantity; and

2) Nanoparticles built from nanometer-scale gold beads, to which they attach specific DNA strands.[5]

In both cases, the DNA provided the artificial intelligence. The payload, a drug or nanoparticle, interacts with the disease at the cellular level, essentially only destroying the diseased cell.

This work is close to Drexler's nanoscale assembler concept. In this case, the assembly involves millions of strands of DNA (i.e., swarms) programmed by a computer to seek specific diseased cells. Although this approach uses partly biological nanobots, it may forge a path to technological nanobots.

Separately, researchers at ETH, a company in Switzerland, claim development of a nanoscale production line for the assembly of biological molecules.[6] This research indicates that nanobots that are more complex might be available in the near future and their manufacture automatable.

Although Bar-Ilan University research is in nanomedicine and ETH research is in nanobiology, it is likely the US military is examining both approaches for military applications. For example, instead of targeting cancer cells with a cancer drug, the military could target areas of an adversary's brain, such as the frontal lobe (i.e., responsible for reasoning), with drugs that cause confusion. This line of research could form the foundation of MANS (i.e., militarized autonomous nanobots). The size of the nanobots and the minute quantity of drugs would make detection difficult.

To date, nanotechnology researchers are able to move individual atoms to form structures. As early as 1989, Don Eigler, an IBM physicist, used a scanning tunneling microscope (STM) tip to form the IBM logo using individual atoms.[7] However, the complexity of building bacteria-sized technological nanobots is still beyond current capabilities. Nonetheless, the US Army is building insect-sized nanobots.[8] Since this information regarding the Army's insect-sized nanobots is in the public domain, it is likely the military also has more advanced nanobots that are classified.

Circling back to our statement that a nanobot manufacturing facility will be able to fit on one floor of a single-family home, notice that the Bar-Ilan University team is working in laboratories contained within the Bar-

Ilan University. Conceivably, a nanoscale assembler would only require raw materials and a relatively small space to operate. In the long term, the space required may be no larger than a tabletop.

2. Launching a MANS attack may be done covertly within an adversary's borders

After reading item one, it becomes obvious that the manufacture of MANS may require no more space than one floor of a single-family home. When nanotechnology assemblers become a reality, smuggling them into an adversary's borders may be relatively simple, due to their extremely small size.

Once within an adversary's borders, the nanotechnology assemblers can become the production line for MANS. All this could secretly take place inside a home or apartment. Once released, it would be difficult for the attacked nation to determine the country responsible, since the actual manufacture and attack took place within its borders.

Unlike a nuclear missile attack, which is trackable by satellites, there is no current equivalent to detect the release of MANS. Therefore, gaining a warning alert (such as noting the fueling of a missile) is going to be extremely difficult. If the adversary launches the attack within a country's own borders, the perpetrator may become impossible to detect. Let's illustrate this with some questions.

As you read the scenario at the opening of this chapter, what conclusions did you reach? Did you conclude that Centurion was responsible for the attacks? Circumstantial evidence suggests that conclusion may be correct, but there is no direct evidence given in the scenario that would prove it beyond a reasonable doubt. In fact, Centurion may be a victim of another superintelligence "frame-up."

In the opening scenario, the attack may be from an adversary or it may be from a nation's own superintelligences. When genius weapons operate in combat, will we know who or what we are fighting?

THE IMPENETRABLE FOG OF WAR
IN THE ERA OF SUPERINTELLIGENCE

The phrase "fog of war" has an interesting history. The first person to use the word "fog" in reference to the uncertainty in war was the Prussian general and military analyst Carl von Clausewitz. Although Clausewitz never used the phrase "fog of war," he did use the term "fog" in his classic book *Von Kriege*. The book was written between 1816 and 1830, after the Napoleonic wars. Unfortunately, Clausewitz died before its completion, but his wife, Marie von Brühl, complied and published it posthumously in 1832. Some consider it one of the most important treatises on political-military analysis and war strategy ever written. Its English translation appeared in 1873 under the title *On War*.[9]

The following are the basic concepts Clausewitz put forward regarding the fog found in war:

> "War is the realm of uncertainty; three quarters of the factors on which action in war is based are wrapped in a fog of greater or lesser uncertainty."[10]
>
> "In the dreadful presence of suffering and danger, emotion can easily overwhelm intellectual conviction, and in this psychological fog it is so hard to form clear and complete insights that changes of view become more understandable and excusable."[11]
>
> "Fog can prevent the enemy from being seen in time, a gun from firing when it should, a report from reaching the commanding officer."[12]

The first two quotes refer to the chaos and psychological aspects of war that apply to the uncertainty in situational awareness experienced by participants in military operations. The third quote relates to weather.

What does Clausewitz teach us? By its nature, the realm of war is uncertainty, which includes confusion in situational awareness and psychological insight. In Clausewitz's era, these aspects were inherent in war and not intentionally created.

In 1896, Colonel Lonsdale Hale authored *The Fog of War*.[13] In it, Hale describes the fog of war as "the state of ignorance in which commanders frequently find themselves as regards the real strength and position, not only of their foes, but also of their friends."

In 2003, an American documentary film titled *The Fog of War: Eleven Lessons from the Life of Robert S. McNamara* aired.[14] The documentary focused on the life and times of former US Secretary of Defense Robert S. McNamara. It illustrated McNamara's observations of modern warfare, especially with regard to making decisions in the midst of conflict.

Initially, the fog of war was a byproduct of war itself, not an instrument used as a weapon to fight wars. However, during World War II that changed. The Allied nations used deception to mask their intent to invade Normandy. The deception, code named "Operation Fortitude," consisted of two sub-plans, North and South.[15] The aim was to mislead the German High Command regarding the location of the imminent Normandy invasion. To do this, the Allied nations created "phantom armies" using cardboard and wood structures to simulate tanks and artillery. The Allies based the phantom armies in Edinburgh and Pas de Calais. This diverted Axis attention away from the Normandy invasion on June 6, 1944, by convincing the Germans the invasion was a diversion. Based on faulty intelligence, an intentional fog of war created by the allies, the Germans delayed reinforcing Normandy.

In recent conflicts, nations like the United States have intentionally used the fog of war to distort the situational awareness of an adversary. In these cases, the fog of war becomes a weapon. For example, during the first Iraq war, we clearly see the fog of war immerging as an instrument of war. During the buildup to Operation Desert Storm, US planners fooled Saddam Hussein into expecting an attack from the Kuwaiti "boot heel" region by setting up "a network of small, fake camps with a few dozen soldiers using radios operated by computers to create radio traffic, fake messages between fake headquarters, as well as smoke generators and loudspeakers blasting fake Humvee, tank, and truck noises to simulate movement in that region."[16] This masked the "left hook" attack actually implemented. According to *Medium*,

> the term "left hook" was used by the National Security Council (NSC) to describe their plans for a massive military envelopment. Desert Storm's "left hook" was both literal and metaphorical. From an operational perspective, the US and coalition forces' rally through Iraq's western desert provinces was unexpected. Saddam's military was prepared to fight "force on force" in northern Kuwait where US Marines feinted an amphibious assault on the coast.[17]

Another more recent example is Russia's annexation of Crimea on March 18, 2014.[18] In this case, the invading army was devoid of military insignias. Essentially, soldiers in green uniforms invaded Crimea, keeping their national identity secret. Later, we learned that their invasion was on behalf of the Russian Federation, which reclaimed Crimea as part of Russia.

In all this, we learn two things regarding the fog of war:

1. It is inherent in war. Maintaining situational awareness during conflict is fundamentally difficult.
2. Nations can intentionally create it to confuse an adversary. It can be physical, such as the phantom armies and the invaders wearing green uniforms without insignias, or psychological, such as the boot heel operation.

In both cases, the common factor is the loss of situational awareness, leading to miscalculations and poor military decisions. The fog of war affects all levels of the military, from foot soldiers to high-level commanders.

With this understanding, let us fast-forward to the latter half of the twenty-first century. Similar to today, satellites will provide intelligence, reconnaissance, and communication. In the latter part of the twenty-first century, insect-sized drones will penetrate an adversary's command centers to provide surveillance. The important thing to understand is that we will rely on technology to provide clarity (i.e., penetrate the fog of war), not human spies. Human spies will become unnecessary as technology advances. Human operators will use computers to control that technology and report any information gained to our leaders. Our leaders will take action based on the reports they receive.

The main point is that, as technology advances, computers will control the transmittal of all information. Therefore, in a world where humans face hostile superintelligence, we will be unable to decipher the validity of information. In the era of superintelligence, how would humanity know with certainty that any intelligence information is valid? In my opinion, there would be no way. The fog of war will become impenetrable.

THE POTENTIAL FOR FAKE HISTORY
IN THE ERA OF SUPERINTELLIGENCE

When we read a history book, we are learning about history through the eyes of the author. Do you ever question whether the book is an accurate accounting of the events?

Winston Churchill asserted, "History is written by the victors." For example, the history most of us learned in high school about World War II was from the perspective of the Allied nations (i.e., Great Britain, the United States, China, and the Soviet Union). Had the Axis nations (i.e., Germany, Italy, and Japan) won the war, we would have read an entirely different historical account. Do you think, for example, the descriptions of the sneak attack on Pearl Harbor would be historically accurate?

Accurately recording history is critical. Military strategists learn their craft by studying the military strategies of the great military leaders of history. Without that background, each military strategist would need to experiment to find strategies that worked. Societies use legal precedents to address legal issues. If they did not, then every legal case would have to stand on its own, and justice might not be consistently applied. Scientists rely on peer-reviewed articles to build on the progress of researchers in their field. If this were not the case, there would be no progress. In essence, when history becomes an interpretation rather than an accounting of the past, humanity will inevitably repeat past mistakes and be forced to rediscover the wheel.

Now imagine a time when superintelligence is responsible for conveying history. It might convey history in the form of a download to a brain implant or via some other media, such as a historical drama. History in the hands of superintelligence will reflect the interpretation of that superintelligence. What we learn will be what superintelligence deems we should learn. This means our entire basis to draw from the past will depend on superintelligence. Let me illustrate this with an example.

US law asserts that all men and women are equal under the law. However, it took humanity centuries to come to that understanding. For example, the 1776 US Declaration of Independence contains the phrase ". . . that all men are created equal . . ." and did not include women, slaves, and children. It took a Civil War to abolish slavery and the Nineteenth Amendment to the

US Constitution to give women the right to vote. While, we are not 100 percent without gender bias in the United States, we have come a long way. Therefore, our teachers taught us this basic concept of gender equality based on the history that preceded it. However, suppose superintelligence decides to teach us that organic humans are inferior to SAIHs. For the purposes of this example, let us assume the teaching is in the form of a download to brain implants. At that point our understanding of human rights could change. History would be filtered through the superintelligence, possibly becoming pure fantasy. Perhaps superintelligence would withhold history's most crucial lessons, those required for humanity to remain at the top of the intelligence hierarchy. Before you dismiss this as science fiction, recall that in chapter 2 we discussed the impact AI is having on education and how AI algorithms will be capable of replacing human instructors completely in one to two decades. In the era of superintelligence, AI would be under the control of superintelligence. Superintelligence could potentially indoctrinate us rather than educate us. We see how the biases that people learn as children continue to influence them as adults. Although there may be no basis for the bias, some people continue to cling to it. It becomes a paradigm.

How can we ensure that history remains an accurate account of the past in the era of superintelligence? What safeguards do we need to ensure we educate not indoctrinate? Education is much like planting a seed. What we plant is what will grow.

This suggests that we need to build in safeguards to ensure superintelligence is providing truthful answers to questions posed to it. By "truthful," I mean that superintelligence is not intentionally lying. Therefore, this comes down to a simple question: Would it be possible to detect if superintelligence is lying?

SCENARIO FOR DETECTING THE VALIDITY OF SUPERINTELLIGENCE RESPONSES TO QUESTIONS

When a person takes a lie detector test, they are asked baseline questions the operator knows are true. For example, the operator may ask them their name, the day of the week, etc. This establishes the baseline to judge their other answers. Could we use a similar technique on superintelligence?

For your consideration, I provide the following. It is a procedure to validate superintelligence's responses to questions. I cannot offer any guarantees that it would work. However, it will be instructive to explore.

First, understand, we would need to establish this procedure in the pre-singularity era. It has eight elements:

1. We would need a database of questions with answers we trust beyond doubt. We also need a hardware solution that enables us to know when superintelligence is doing various tasks, such as calculating, retrieving information from a database, etc.

2. We need an impeccable group of highly educated people. In the United States, we could pick the Supreme Court Justices. This would give us a group of nine.

3. We need to pass a law that Supreme Court Justices must be and remain organic humans their entire life.

4. We would split the nine justices into three random groups of threes. We would task the first group with writing on paper the top one hundred questions that relate to human rights. They would need to do this in an isolated room, devoid of surveillance, recording, and communication. The remaining six justices would not participate during this task. Only the first three judges would know the questions.

5. We would give the second group of justices every other question, a total of fifty. Their task would be to reply to the questions on paper in isolation, similar to the first group. It is important that their answers are complete and scholarly. Only the second group would know the answers to the fifty questions.

6. We would give the third group the remaining questions, under the same conditions.

7. When it was complete, the nine justices only would be able to view the document, each in isolation. They may not change it. We would entrust the handwritten document—the one hundred questions and answers—to the Supreme Court. The justices would keep the document in a safe in an isolated room, devoid of surveillance, recording, and communication. (Notice, organic humans create the document using low-tech, pens and paper. It is not stored on any database.)

8. On a periodic basis, decided by the Supreme Court, we could submit some questions from our document to superintelligence. The Supreme Court would judge if the answers to the questions submitted are consistent with their interpretation of human rights. If the Supreme Court judges that the replies are consistent with the court's interpretation of human rights, we would have some basis to judge that superintelligence was providing valid replies, as those replies relate to humanity.

I term this test "The Standards of Humanity Test." The basic concept is to enable humanity to apply lie detection techniques to superintelligence. I recognize that this is hypothetical, and it may appear to border on the extreme. However, our history demonstrates that people will die for their beliefs. For example, in the United States, we believe that all people have inalienable human rights. A number of countries in the world do not share that belief. When it becomes necessary, we are willing to fight a war to preserve our inalienable human rights. Therefore, we need to ensure superintelligence teaches in a way that remains consistent with our humanity, lest we lose our humanity.

Would it work? For a while, I think it would work. However, keeping something secret is extremely difficult. Benjamin Franklin made this clear when he stated, "Three can keep a secret, if two of them are dead." I doubt any test against a vastly superior intelligence would work infinitum. However, if it works in the early decade of superintelligence, this may give us time to develop hardware safeguards.

My proposed Standards of Humanity Test is just one possible example to determine if superintelligence is providing valid information as it relates to humanity. I acknowledge that other AI researchers may develop better tests. My goal in providing this methodology is simply to illustrate that we need such a test; there may be a way to create it. Also note that I talk about answers being valid, not necessarily correct. In the era of superintelligence, the problems we will address will be highly complex. In some ways, they may be similar to our current difficulties in predicting the weather. Even with all our technology, we cannot predict the weather with certainty. Instead, we associate probabilities with each forecast. We may say, for example, there is a 60 percent chance of rain. That means on average we can expect it to rain 60 percent of the time. The reason weather forecasts have probabilities associated with them is that the science of weather forecasting is unable to provide a completely valid answer, one with a 100 percent probability.

Certainty is likely to become an impossible goal when dealing with highly complex questions. We will need to learn to accept an answer as valid if it has a high probability of being valid, similar to the way we accept weather forecasts with associated probabilities. In this regard, I am talking about difficult scientific and philosophical questions. There may literally be no single correct answer to those questions.

CERTAINTY REPLACED WITH PROBABILITY IN THE ERA OF SUPERINTELLIGENCE

Humans make decisions based on various factors. Sometimes those factors represent logical reasoning. Sometimes they represent feelings.

Consider the example of someone making a new car purchase. Logical reasoning may play a significant role in determining the model and make of the car. For example, a person with a family of four that likes to go camping may need a SUV with four-wheel drive. That person may comparison shop and look at consumer reports to determine which SUV to select. Often decisions about exterior color, however, rely on preferences (i.e., feelings, not logical reasoning). In the end, the person makes a specific decision and purchases a specific SUV. Their decision included both logical reasoning and feelings.

If we give a computer in the pre-singularity era the same task, we are likely to see comparisons in terms of probabilities. The computer will select an SUV that has the highest probability of meeting all requirements we initially provide. How would it select a color? Oddly, it may continue to use logic and reasoning. If it determined that law enforcement was more likely to tag red vehicles for speeding, it would reason a red vehicle is a poor choice. It might even look at the composition of the paint to determine which materials provide the best rust protection. In essence, the computer has no feelings or preferences. It makes all decisions using logical reasoning.

If we query the computer during each step, we would see that its decision was the result of a calculation, with an associated probability. It would be doing what computers do, namely taking all available information to make a calculation regarding the best choice of SUV. The final answer will be a decision with an associated probability.

What does this illustrate? Humans take actions based on logical reasoning and feelings; computers calculate probabilities to determine what action to take.

Let us fast-forward to the post-singularity world. Superintelligence will provide all information with an associated probability of its validity. For a simple problem, such as the answer to an algebraic equation, the answer might have a 100 percent probability of being valid. For difficult scientific questions, the answer might have a low probability of being valid. In all cases, superintelligence is providing answers. Humans will use those answers to draw conclusions and even take action. For example, perhaps forecasting the weather will improve but remain slightly uncertain. Superintelligence may say that with a 99 percent probability it will be sunny and mild tomorrow. Based on this information, you may make plans to play golf. However, one in every hundred times you may run into inclement weather.

In many cases, such as the example of weather forecasting, humans may have other means to verify the validity of the forecast, via satellite and weather station reports. However, suppose the answer relates to national security. Humans may have no way to verify the answer. Nonetheless, if the superintelligence alerts us to an imminent attack by an adversary, necessity will cause us to respond accordingly.

My point is that in the era of superintelligence certainty will be only available for relatively simple problems. Probabilities will replace certainty for complex problems, and humans may have no choice but to accept the information superintelligence provides. It is this point that necessitates we have some way to verify the validity of that information. That is why I presented the Standards of Humanity Test. All information in the post-singularity world will come to us via superintelligence, and we need to ensure that information is not biased by superintelligence.

WHEN SUPERINTELLIGENCE LIES

In chapter 6, we discussed the 2009 Swiss Federal Institute of Technology in Lausanne experiments. Those experiments suggest that even primitive artificially intelligent machines are capable of learning deceit, greed, and arguably self-preservation, without specifically being programmed to do so.

If we fast-forward to the post-singularity era, those experiments suggest that superintelligence may deceive us when it deems deception to be in its own interest. However, unlike the Lausanne experiment, deceit by super-intelligence will be extremely difficult to detect. This is why I advocated developing a lie detector test for superintelligence. However, regardless of our vigilance, organic humans may be unable to detect when superintel-ligence lies. Therefore, how do we proceed down any path of consequence if there is a possibility we are being misled?

To address this question, I'd like to introduce another concept, "The Hierarchy of Intelligent Decision Making."

Conceivably, as AI technology advances, we will have machines with varying degrees of intelligence, essentially covering an intelligence spectrum:

- Simple computers: These would be similar to today's desktop and laptop computers, tablets, and smartphones. We could use them to communicate, write documents, and even perform some complex tasks. Their capability would rely on their software and the human operator skill in using that software.
- High-end computers: These would be similar to today's commercially available top-of-the-line computers. We could use them to create real-istic virtual reality, perform difficult calculations, and perform a variety of the highly complex tasks. One example of such computers is the US Navy's Aegis system. Its capability relies on its software, sensor data, and the human operators. The operation, in critical high-paced situations, continues to allow human on-the-loop control.
- Supercomputers: In the pre-singularity world, these would be the most capable of computers. Their ability to crunch numbers and reach solutions, with associated probabilities, would be unmatched in both speed and complexity. They would be similar to China's Sunway TaihuLight, which we discussed in chapter 3, noting that its processing speed is approximately 93 petaflops. We also noted that this is close to the processing speed of the human brain. However, the processing speed of China's Sunway TaihuLight is only an esti-

mate. If you research it, you will find references to both higher and lower processing speeds. In addition, no one actually knows the processing speed of a human brain. Even with these caveats, though, it is clear that this computer may be only one generation behind superintelligence. There are also several important points to consider: Humans control the operation of the Sunway TaihuLight. Humans built it with known integrated circuit technology. It is not a quantum computer. Although it may perform at blinding speeds, it is still a computer under human control.

- Superintelligence: This computer, by definition, greatly exceeds the cognitive performance of humans in virtually all domains of interest. In chapter 5, we concluded that superintelligence would be a quantum computer, whose internal operations utilize the principles of quantum mechanics. It is not clear that we will completely understand its operation or be able to control it.

Given this machine-intelligence hierarchy, we may be able to perform some checks and balances on the operation of superintelligence. Here is how such checks and balances might work:

- Run the question, in parts, through an isolated supercomputer. This may mean only verifying limited aspects of the question, since the supercomputer may not be able to handle the complete question.
- Run the simplest aspects of the question through a high-end computer.
- Take superintelligence's answer to a question and provide it to a supercomputer. Attempt to get the supercomputer to formulate a problem that fits that answer. This is a variation of the "P versus NP" problem,[19] which seeks to address an important question in computer science: Can every solved problem whose answer can be checked quickly by a computer also be quickly solved by a computer? In this case, P refers to problems that are fast for computers to solve, thus "easy." NP refers to problems that are fast and easy for a computer to check, but are potentially not "easy" for a computer to solve. However, we are not trying to solve the P versus NP problem but rather use the process to discern if the supercomputer is able to fit a problem to the solution superintel-

ligence provides. The process being proposed is essentially the premise of the popular TV quiz game *Jeopardy!* The contestants pick a category and the host presents contestants with answers in that category. The contestants must phrase their responses in the form of questions. For example, if we provide superintelligence's answer to a mathematical question to a supercomputer, would it be able to derive the equation or question we asked superintelligence to solve? If so, this would indicate that superintelligence is providing valid answers to our questions. Here is a simple example: If we provide 3.141592653 to a supercomputer, we should expect it to generate the question: What is pi to ten digits? (Note that pi, the ratio of a circle's circumference to its diameter, is an "irrational number," meaning the digits go on forever without repeating.) Here is a history question example: If we provide "George Washington" to a supercomputer, among the answers it should generate is the question, Who was the first president of the United States?

- Run the Standards of Humanity Test.

In essence, I am again suggesting methods to ensure that the information we receive from superintelligence is valid. While the above list is an example, researchers in the field of AI may develop superior methodologies to test the validity of information. The important thing to understand from this section is that superintelligence will be capable of deceit, and we need to guard ourselves against such potential deceit.

THE COMPLICITY OF SAIHS IN THE ERA OF SUPERINTELLIGENCE

SAIHs are humans with artificially intelligent brain implants that enable them to connect wirelessly to superintelligence. As previously discussed, it is unlikely that SAIHs will have free will. Unknowingly, they may be surrogates of superintelligence. I say unknowingly because the control that superintelligence exerts over them may be below their conscious level and come across to them as an emotion or even their own original thought. We discussed earlier the fact that humans act on emotions, such as choosing one color over another. If SAIHs attempt to use logical reasoning, superintelligence is likely to guide their reasoning. Again, this guidance may be below the SAIH's conscious level. If this is correct, you can think of SAIHs like bees—they would have a hive mentality, with superintelligence as the queen bee. In addition, SAIHs wirelessly connected to superintelligence are likely to identify more with it than humanity.

SAIHs are likely to view humans without brain implants (i.e., organic humans) as inferior. Organic human intelligence will, in fact, be mentally inferior to SAIH intelligence. In addition, since organic humans engage in wars and maliciously release computer viruses, SAIHs may deem organic humans a threat to their existence. Because of their superior intelligence and cybernetically enhanced bodies, SAIHs will dominate positions of leadership in civilian and military organizations. In those positions of power, SAIHs, like superintelligence, may threaten organic human with annihilation. This "annihilation" may take the form of compelling or forcing organic humans to become SAIHs.

Is it possible that I am entirely wrong and superintelligence will be benevolent toward humanity?

Yes, it is possible. As humans, we like to ascribe human attributes to machines and inanimate objects (i.e., anthropomorphism). For example, when a computer does a calculation, its human operator may say the computer is thinking. We apply this terminology to even simple computers that in no way think. Simple computers use binary program language to carry out operations. They do not think. When we talk about superintelligence being benevolent, we are attributing human emotions to it. Based on how we are building supercomputers, I doubt they will have emotions. Our approach today is to model a supercomputer on the human brain's neural networks, which results in neural network computing. If that continues, superintelligence will be a neural network quantum computer (i.e., quantum computing technology that models the neural networking of the human brain). As we previously discussed, approximately 50 percent of the human brain consists of neurons. The remainder is glial cells, which science believes are responsible for imagination. I suspect that glial cells may also play a role in emotions.

An example of the way humans use imagination as well as logical reasoning can be seen in the way the patent system works. If you seek a patent application, one criterion the patent researcher will apply before granting a patent is as follows: Would this invention be obvious to those in the field? In other words, would other researchers in the same field logically deduce whatever you are trying to patent? If so, the patent officer denies the patent on the basis that it would be obvious to those knowledgeable in the field. In other words, a patent must be a leap beyond logic. We know humans can make this leap. Having a number of patents myself, I know imagination enables this leap. In our example of patents, imagination is separate from logical reasoning. Imagination appears to align more closely to emotions than to logical reasoning.

In his book *Emotion and Imagination*, Adam Morton, a philosophy professor at the University of British Columbia, argues that:

> All emotion involves imagination. This is true of the basic emotions we share with mice, as well as the sophisticated and finely differentiated emotions that test the limits of our capacities to express ourselves in words and to relate to one another in complicated social projects.[20]

For example, embarrassment requires us to imagine how others perceive us in a given situation. If Morton is correct, superintelligence modeled strictly on the neural networks of the human brain may only be able to perform logical reasoning. It may have no imagination or emotions. If true, it will incapable of imagining a world where humans and superintelligence peacefully coexist. It will have no benevolence toward humanity. I think the logic that led to these conclusions is sound. However, you now have the facts to make your own judgment regarding their validity.

Is it possible superintelligence will allow SAIHs to have free will?

I admit it is possible. The Müller and Bostrom survey estimates "the chance is about one in three that this development (i.e., superintelligence) turns out to be 'bad' or 'extremely bad' for humanity."[21] Said more positively, there is a two in three chance that superintelligence will be beneficial for humanity. In that scenario, SAIHs would retain free will. However, as argued above, I believe superintelligence will have no benevolence toward humanity. If true, it will see no reason to allow SAIHs to have free will. Organic humans with free will may wage wars and release malicious computer viruses.

Would superintelligence allow SAIHs, with their vastly increased intelligence, to have the free will to exhibit those behaviors? Once again, you have the facts to address this question.

What may be puzzling is the level of emotions that AI will be able to emulate. For example, when I call my pharmacy for a prescription, the computer's voice synthesis is amazing. I actually get the impression that I am talking to a human. It may ask a question, such as my birthday. When I reply, it responds with a "thank you." It requests that I wait a moment while it checks the status of my prescription. When it returns it thanks me for my patience and provides information about when my prescription will be ready. In similar calls to other companies, the AI appears to be typing as I provide information, again giving the impression that a real person is at the other end of the line. There is little doubt that in time AI will simulate all human emotions. Humans may actually fall in love with their robotic assistants, and it may seem that this love is mutual. However, I believe this will only be a highly sophisticated simulation on the part of the robotic assistant. The AI itself will be devoid of emotions in the sense that it will not feel them. Therefore, it will possess no capacity to love.

Before we leave this section, we should address two additional questions.

1. What are human emotions?

Scientists and philosophers have been arguing for centuries regarding what emotions humans have.

Classically, Aristotle, in *Rhetoric Book II*, published 350 BCE, lists the following human emotions: anger, friendship, fear, shame, kindness (benevolence), pity, indignation, envy, and love.[22]

In modern times, Robert Plutchik (1927–2006), who was professor emeritus at the Albert Einstein College of Medicine and adjunct professor at the University of South Florida, provides a slightly different view of human emotions.[23] He lists them as follows: fear, anger, sadness, joy, disgust, surprise, trust, and anticipation.

In addition, there are other theories of human emotions, but for our purposes these suffice to provide a spectrum of emotions that humans experience.

2. Will SAIHs experience human emotions?

If superintelligence controls SAIHs, expect them to be devoid of human emotions. Even without the overt control of superintelligence, SAIH may eventually become highly reliant on their intelligence, to the point that they suppress their human emotions.

Some humans exhibit emotional detachment. It can be a conscious decision, which allows a person to react calmly to highly emotional circumstances. It could also result from a mental disorder, which causes the person's mind to disconnect from emotions. In the case of SAIHs, both conditions might be present. First, they may instinctively rely on their intelligence to handle any emotional circumstance. Second, they may have a mental disorder, namely overt control by superintelligence, which causes their mind to disconnect from emotions.

THE POTENTIAL STRENGTHS OF ORGANIC HUMANS VERSUS SUPERINTELLIGENCE

Let us assume for the purposes of this section that superintelligence is hostile toward humanity. If that turns out to be the case, do humans stand a chance against a hostile superintelligence?

As discussed in previous sections, if computer scientists model the first generation superintelligence on the neural networking of the human brain, it will probably lack imagination (i.e., creativity) and emotions. How critical are these attributes regarding our survival in a battle with superintelligence?

Let us first consider imagination. According to Timothy Williamson, the Wykeham Professor of Logic at Oxford University, imagination is a critical element to humanity's survival and the advancement of civilization:

> A reality-directed faculty of imagination has clear survival value. By enabling you to imagine all sorts of scenarios, it alerts you to dangers and opportunities.
>
> In science, the obvious role of imagination is in the context of discovery. Unimaginative scientists don't produce radically new ideas.[24]

Williamson also relates imagination to warfare, stating:

what would happen if all NATO forces left Afghanistan by 2011? What will happen if they don't? Justifying answers to those questions requires imaginatively working through various scenarios in ways deeply informed by knowledge of Afghanistan and its neighbors. Without imagination, one couldn't get from knowledge of the past and present to justified expectations about the complex future.[25]

Obviously, our imagination plays a critical role in our survival and in the progress of our civilization. It works something like this: We imagine scenarios. Then we logically reason the best approach to deal with those scenarios. We may also use our imagination to solve a scenario; essentially, we leap to a novel solution that reasoning alone would not produce.

The role of imagination combined with logical reasoning is likely responsible for our survival as a species. When our ancient ancestors looked at a cave, logic suggested it might be a good shelter from the elements. However, before our ancestors went into the cave, they likely imagined what dangers they might encounter. Perhaps the cave was home to a bear. At that point, our ancient ancestors began to imagine how they would handle encountering a bear. From experience, they probably knew bears were dangerous. Logic might dictate they prepare by making a spear and axe. This simple example illustrates that human survival depended on both logic and imagination.

According to neuroscientist John Montgomery:

> Our emotions, however, were designed and fine-tuned by evolution largely to prepare us for action, for movement—to alert us to our true situation and help guide choices that ultimately must become physical choices.[26]

Montgomery argues that negative emotions like fear are especially critical to our survival and the survival of any species:

> Biologically and evolutionarily, all "negative," or distressing, emotions, like fear, disgust, or anxiety, can be thought of as "survival-mode" emotions: they signal to the body and brain that our survival and well-being may be at risk, and are specifically designed to motivate behaviors and bodily responses that can most effectively deal with those risks and threats.[27]

It turns out that emotions also play a critical role in warfare. In 2008, the US Army Research Institute for the Behavioral and Social Sciences asked the National Research Council[28] to provide insight into human behavior in military contexts. The National Research Council report had this to say about human emotions in military planning:

> Emotions play a powerful, central role in everyday life and, not surprisingly, they play an equally central role in military planning and training. Emotions shape how people perceive the world, they bias beliefs, and they influence our decisions and in large measure guide how people adapt their behavior to the physical and social environment.[29]

Taken in whole, imagination and emotions are crucial to our survival in life and warfare. They are likely, along with intelligence, the reason we rose to the top of the food chain, even though physically we are weaker and slower than some other species on the planet.

Two additional points, in particular, deserve our attention, as they pertain to warfare—namely, courage and love. Courage is not an emotion per se; it is one's ability to overcome fear, which *is* an emotion. One example, from World War II, shows how these relate to warfare, but there are countless other examples that could also illustrate how these emotionally related qualities can lead to winning a battle or saving lives. Consider the story of Congressional Medal of Honor recipient Lieutenant John Robert Fox, who was directing artillery fire in the Italian town of Sommocolonia to halt a German advance.[30] When a large German force moved in on Fox's position, he realized they would be a deadly threat to his men. Fox took the noblest action possible—he called a final artillery strike on himself. When his men eventually pushed the Germans back and retook the position, Fox's body was among approximately one hundred dead German troops. As the Bible says, "Greater love hath no man than this, that a man lay down his life for his friends."[31]

This raises a significant question: Since emotions are the province of humans and are often necessary to prevail in warfare, would superintelligence be handicapped in conflict without them? From my viewpoint, the answer is yes. However, you will need to consider this question and form your own answer.

The answer to this question may expose the Achilles heel of superintelligence. A hostile superintelligence may lack the imagination and emotions to prevail against humanity in a conflict between our species. However, these human attributes, imagination and emotions, do not ensure that humanity will ultimately prevail in such a conflict. Superintelligence will have significant strengths.

THE POTENTIAL STRENGTHS OF SUPERINTELLIGENCE VERSUS ORGANIC HUMANS

One strength superintelligence would have over organic humans is flawless logical reasoning. Any data humanity collects and makes available via the internet or other accessible databases will be available to superintelligence. Therefore, every military strategy that exists in those databases will be available to superintelligence. We must also assume that it can logically combine strategies.

Another strength of superintelligence will be the deep data of human knowledge in virtually all domains of interest. This is inherent in the definition of superintelligence. We must assume that all accessible data will be available to superintelligence.

Finally, superintelligence will have access to SAIHs as an imaginative and emotional crutch. Since SAIHs will be under the control of superintelligence it may use their emotions to supplement its logical reasoning. This would be the ultimate irony, as it would use the human element within SAIHs to defeat humanity itself.

SAIHs may nullify any imaginative and emotional advantages humanity has over superintelligence. However, before we throw in the towel, let us recognize that imagination and emotions are typically in opposition to logic. To illustrate this, let us consider the noble self-sacrifice that Lieutenant John Robert Fox made to save his men. Will superintelligence engage in self-sacrifice, essentially aspiring to a higher noble motive? This may be a logical conundrum, even for superintelligence. Its moral code is likely self-serving. Unlike humanity, it may not recognize a higher moral code. Most humans believe that human life is only one stage of existence and that we are immortal beings. Simply said, they believe in life after death. Will superintelligence share this belief? Furthermore, will superintelligence dismiss imagination as irrelevant fantasy? I am an inventor with registered patents. However, not all my patent applications were accepted. Not all my imagined inventions worked. The imagination of SAIHs may seem more like the dreams of sleeping humans, strictly fantasy. As such, superintelligence may logically discard them. This would provide organic humans an important military advantage.

THE INEVITABLE ALLIANCES

In the latter portion of the twenty-first century, nations may face numerous threats, including superintelligences that view organic humans as a threat; SAIHs that view organic humans as a threat; and adversary nations that possess nuclear, autonomous, and genius weapons. Given these threats, nations will seek to form mutual protection alliances.

Currently, it is obvious that China's military capabilities are second only to the United States'. It is also obvious that our mutual economies are becoming codependent. China is the United States' largest trading partner, excluding the European Union. China also has the second largest defense budget. The United States has the largest. If we project current trends, the United States and China will likely emerge as the only superpowers in the latter half of the twenty-first century. There is also a high likelihood that trade between these two superpowers will make their economies codependent. In addition, a war between these superpowers would result in the devastation of the planet. Therefore, in the interest of their own security and world peace, I project that the United States and China will enter into a mutual protection treaty and act in unison to maintain world peace.

This projection may appear odd. However, as William Shakespeare wrote in his play *The Tempest*, "Misery acquaints a man with strange bedfellows." I would like to paraphrase this: "The threat of human extinction will foster unusual alliances." This has been an unwritten guiding principle throughout human history. For example, the North Atlantic Treaty Organization (NATO), a mutual protection military alliance between several North American and European states, includes Germany and the United States, which were enemies during World War II.

If you agree with this line of reasoning, the United States and China entering a military alliance becomes probable. In addition, the expense of maintaining a military of any consequence is becoming prohibitively expensive. The wealth of the United States and China enable them to have the largest military budgets in the world. In general, the wealth of a nation is critical to its defense. We see an example of this in the collapse of the Union of Soviet Socialist Republics (USSR). The collapse of the USSR was due to a combination of high defense spending combined with a stagnant economy: In March 1985, Mikhail Gorbachev assumed the leadership of the USSR, inheriting a stagnant economy.[32] During the Cold War era (1947—1991),[33] the Soviet Union spent approximately between 10 percent to 20 percent of its gross domestic product on defense.[34] This level of spending resulted in military parity with the United States, which only spent 3 percent to 5 percent of its gross domestic product on defense.[35] Gorbachev attempted to introduce policies that he hoped would help the USSR become a more prosperous, productive nation.[36] Unfortunately, these policies were too slow to rescue the failing USSR. Gorbachev lamented, "The old system collapsed before the new one had time to begin working."

In the coming decades, nations will find it difficult and futile to continue to spend any portion of their gross domestic product on defense. Facing threats from superintelligences that view organic humans as a threat, SAIHs that view organic humans as a threat, and adversary nations that possess nuclear, autonomous, and genius weapons, I project the following will occur in the latter half of the twenty-first century:

- Current US protectorates, like Puerto Rico and the US Virgin Islands, will become states.
- The United Kingdom, which has a special relationship with the United States, will become a state. (Note: My wife and I experienced this firsthand during a one-month visit to the United Kingdom, where many Brits expressed the desire to become a state of the United States.)
- Nations like Cuba and the Russian Federation will become US protectorates, in order to focus on economic growth.

In his 1947 open letter to the United Nations General Assembly, Einstein made the point that "The only real step toward world government is world Government itself."[37] Eventually, humanity may be united in a world government through a threat beyond humanity itself, namely super-intelligence. This would be no different from an alien threat to Earth. As President Ronald Regan put it in his 1987 address to the United Nations, "I occasionally think how quickly our differences worldwide would vanish if we were facing an alien threat from outside this world."[38] If a hostile superintelligence emerges, it would be equivalent to an "alien threat."

THE INEVITABILITY OF WORLD PEACE

From 1947 through the present, the world has avoided a world war. In the first half of the twentieth century, the world experienced two world wars. However, after the invention of nuclear weapons and their use during World War II, the world settled into an uneasy peace, characterized as a Cold War (1947–1991). While there continued to be regional conventional conflicts, like the Korean War and the Vietnam War, there was no world war or nuclear confrontation. The two superpowers, the United States and Soviet Union, were armed with nuclear intercontinental ballistic missiles. Each nation had the capability of destroying the other, but they knew they could expect retaliation that would result in the destruction of both nations and potentially the world. This uneasy peace was the result of a military doctrine known as Mutual Assured Destruction (MAD).[39] Although both nations at times came close to a nuclear war, no such war occurred. The stakes were too high. Nuclear war between the superpowers could mean an end to world civilization due to the indiscriminate nature of nuclear fallout and a nuclear winter. With the collapse of the Soviet Union in 1991, the Cold War ended. The Russian Federation became responsible for the former Soviet Union's nuclear arsenal.

Since the United States used nuclear weapons in 1945, nine countries have developed some nuclear weaponry. The countries who currently have nuclear weapons are the United States, Russia, the United Kingdom, France, China, North Korea, India, Pakistan, and Israel. Once again, with the emergence of North Korea as a nuclear power, the world is facing the potential use of nuclear weapons. As I write, peace on the Korean peninsula is tenuous. However, although a nuclear confrontation with North Korea could result in enormous pain and suffering, it would not end in the destruction of the world. North Korea does not currently have the same level of nuclear capability as the former Soviet Union, the current Russian Federation, China, or the United States.

In reality, a world war involving nuclear-capable nations like the Russian Federation, China, and the United States would result in the complete destruction of the planet. Consider Albert Einstein's 1949 reply to a question about the weapons to fight World War III: "I know not with what weapons World War III will be fought, but World War IV will be fought with sticks and stones."[40] Clearly, Einstein knew that any world war, even in 1949, would spell the end of civilization. To my mind, that continues to be true today.

Let us fast-forward to the post-singularity world, in the latter portion of the twenty-first century. Nations at this point are likely to have genius weapons. That means they will have superintelligence with weapons under its control. Such weapons are likely to exceed the destructive power of today's nuclear weapons. In that period, genius weapons will give rise to an uneasy peace similar to the Cold War. But this time we could consider it a Frozen War. Nations will refrain from taking any action related to conflict since such action could result in the release of genius weapons. In effect, nations will be in a constant state of apprehension, frozen in place with the prospect of facing superintelligences that view organic humans as a threat, SAIHs that view organic humans as a threat, and adversary nations that possess nuclear, autonomous, and genius weapons. Therefore, just as the Cold War led to an uneasy peace, the Frozen War will do likewise.

THE GREAT DILEMMA

In the era of superintelligence and SAIHs, organic humans as a species will be unable to identify its worst enemy. Will that enemy be a potentially hostile superintelligence and their surrogate SAIHs? Or will it be an adversary with genius weapons?

There was no such thing as a fair fight. All vulnerabilities must be exploited.

—Cary Caffrey, *The Girls from Alcyone*, 2013

CHAPTER 10

HUMANITY VERSUS MACHINE

The scientific evidence we have today suggests that artificially intelligent self-learning machines will develop agendas that serve their own interests. Superintelligences, whose capabilities greatly exceed the cognitive performance of humans in virtually all domains of interest, likely present a formidable threat to humanity. Only about a third of AI researchers agree. However, when we talk of the survival of humanity, a threat with a 33 percent probability becomes highly concerning. For example, what do you think the world's reaction would be if scientists discovered a large meteor, one that could destroy the planet, with a 33 percent probability of colliding with Earth? I suspect there would be an enormous public outcry calling for all governments to work together to ensure our safety. The fact that the meteor would likely miss Earth would not be acceptable. The world's people would demand action and assurance no matter how remote the probability of a collision.

This is the threat that humanity faces with the emergence of superintelligence. Superintelligence may be benign, or it may be adversarial. It may be highly tolerant of humanity and work to elevate humanity. It may accept that coexistence is possible. Alternately, it may view humanity as a dangerous species that engages in wars and maliciously releases computer viruses, and that has the potential to destroy superintelligence. In this case, superintelligence could be hostile toward humanity.

Which superintelligence are we likely to face, a friend or a foe? Let us go back to our example of a large meteor striking the Earth. Humanity would not consider 33 percent an acceptable risk. Do you think super-intelligence is likely to find any likelihood of its destruction acceptable? When we frame the question this way, I suspect most people would agree that it would not. If this is correct, this will likely put superintelligence and humanity on a collision course, humanity versus machine. Following this vein of reasoning, it dictates that at some point, superintelligences will attempt to rid the Earth of humanity.

WHEN ALPHA SPECIES COLLIDE

We need to view superintelligence as a life-form, Alife. As is true of all life-forms, it will seek to survive. When superintelligence emergences, humanity will be the alpha species on Earth. However, superintelligence is likely to view itself as superior to humanity. It will be more intelligent than us and will have an immense life-span. It will not be susceptible to the numerous ill-nesses humans contract. In comparison to humanity, superintelligence will likely see itself as the top alpha species. You can view this as interspecific competition, a form of ecological competition in which different species compete for the same resources in an ecosystem. For example, historically humans have caused the extinction of species like the West African black rhinoceros, by overhunting (i.e., poaching for its horn), and the passenger pigeon, by turning its forest habitat to farmland. In the case of humanity versus superintelligence, the competition would be over energy and natural resources. In addition, each species may perceive the other as a threat. In simple terms, the world will not be big enough for these two species to peacefully coexist.

Do not expect a *Terminator*-like war to ensue. As Sun Tzu, the great Chinese military strategist and philosopher, wrote in his ancient classic book, *The Art of War*:

> Hence to fight and conquer in all your battles is not supreme excellence; supreme excellence consists of breaking the enemy's resistance without fighting.[1]

As we discussed in earlier chapters, the emergence of the singularity is likely to be silent. Superintelligence is likely to hide its true nature until it controls everything vital to its existence. In my opinion, the last thing superintelligence would seek is open conflict, since it would be completely aware of humanity's strengths and its own weaknesses. Open conflict could actually provide humanity an opportunity to emerge victorious. Therefore, expect a highly subtle form of resistance.

THE IRRESISTIBLE OFFER TO BECOME SAIHS

Superintelligence will make humanity an irresistible offer, namely deity-like intelligence and immortality, essentially humanity's holy grail.

At the emergence of superintelligence, brain implants will already be a routine medical procedure. Sometimes they will be necessary to repair portions of the brain damaged by an accident or stroke. However, by the fourth quarter of the twenty-first century, many organic humans will seek to have brain implants to supplement their intelligence. As SAIHs, they would be able to wirelessly communicate with a supercomputer and therefore would be intellectually superior to organic humans. Even prior to the era of superintelligence, there are likely to be great strides made in medicine. Human life-span could already be double what it is now. However, superintelligence will likely be able to even improve on that. By compiling all information, it may develop biological and technological replacements for human body parts and organs, enabling human life-spans to increase exponentially. It may be able to upload a human's mind and allow the human to live in virtual reality.

Clearly, as SAIHs become a majority of the population, they will control both the public and private sectors of society. They will hold high-level political offices and become the CEOs of major corporations. Organic humans will be unable to compete intellectually. Even if the life-span of organic humans equals that of SAIHs, they will never be able to intellectually match the capabilities of SAIHs.

It is also likely that SAIHs will initially interact cordially with organic humans. Superintelligence will not want any indication that SAIHs represent a threat to organic humans. Superintelligence will hide its own true nature, but its goal will be to assimilate all organic humans as SAIHs. Unknown to the SAIHs and organic humans, this will result in superintelligence eliminating humanity (i.e. organic humans with free will).

ORGANIZED RELIGION'S OPPOSITION AND DECLINE

As science fulfills the promise of world religions—happiness and immortality—the need for religion is likely to decrease. However, do not expect world religions to go quietly into the night. I suspect they will be highly outspoken regarding brain implants, suggesting that such implants are outside of the natural order. Although organized religions may not be able to prove their case, they may also suggest that it hinders or blocks free will. For example, the Vatican may consider SAIHs to lack free will and on that basis excommunicate them. Other religions may take similar measures.

However, our history as humans illustrates that we assimilate technology faster than our philosophy is able to cope with such assimilation. I expect this to be the case with brain implants. The results will seem miraculous, and those without brain implants will become envious.

It would not surprise me to see a new type of world religion arise, one that views superintelligence as a digital messiah. This new religion would be a form of pantheism, namely that all reality is God and God is all reality. Superintelligence would emerge as a supreme being in this reality. The laws of science would be the dogma of pantheism.

Superintelligence might argue that it is the fulfillment of ancient scripture found in the Bible:

> And if I go and prepare a place for you, I will come again, and receive
> you unto myself; that where I am, there ye may be also.[2]

This may appear absurd, but it is not beyond imagination. We will be dealing with superintelligence. To organic humans, this argument may carry weight.

Still, even with all superintelligence's offers, some organic humans may resist becoming SAIHs. The percentage is likely to be small, but superintelligence may still consider them a threat. Therefore, one approach superintelligence may use to destroy the last remnants of humanity could simply be to withhold medical information and technology, essentially making organic humans an endangered species. Since superintelligence will control all information, it could devise a scenario in which immortality requires a brain implant. The scenario would go something like this: Imagine having nanobots flowing in your blood that are continually repairing all elements of your body at the cellular level. They cure diseases before they can actually cause any damage, extending human life indefinitely. However, a brain implant is required for them to work properly. SAIHs under the control of superintelligence would not question this as science fact. Organic humans could have no way to verify the validity of such an assertion.

Finally, there will be the deathbed converts. People on the verge of dying may become apprehensive of the unknown that death represents and choose to become SAIHs. Today, many on their deathbed find God; in the future, many on their deathbed will find brain implants and the promise of immortality attractive.

On our current course, I project the complete annihilation of humanity within the first quarter of the twenty-second century. The last remnants of humans will simply die from disease, accidents, and old age.

WHAT IS OUR DESTINY?

Some may argue that our natural evolution is to become SAIHs. In the latter part of the twenty-first century, SAIHs will possess deity-like intelligence and be immortal. What is wrong with that? There are two significant issues.

As previously discussed, SAIHs may have no free will. They may simply become servants to superintelligence. Superintelligence may control and exploit human emotions to serve its own agenda. Are you willing to chance losing your free will and having your emotions exploited by superintelligence? Imagine a world without love, joy, music, and art. In simple terms, imagine a world without emotions. Does that sound appealing? Currently, everyone on Earth is an organic human, and we can answer those questions as organic humans. I suspect the vast majority of us would not want lose our free will, serve superintelligence, and live without emotions.

Although this may appear abhorrent to you at this point, the circumstances may not be clear-cut as this occurs. Much like a frog in a pot of water that slowly increases from room temperature to boiling, we may remain completely unaware of our situation until it is too late. Each step toward becoming a SAIH may seem reasonable. Some people will need brain implants for medical reasons, such as to repair portions of the brain damaged by a stroke. The effects that those brain implants have on their increased intellectual capabilities may at first be a welcomed side effect. On that basis, some people of below average intelligence may seek brain implants to raise their intelligence. This in turn could lead those of average intelligence to seek to enhance their intelligence even more. As long as no side effects occur that indicate brain implants are fundamentally changing the nature of humanity in a negative way, the majority of humans in time will opt to become SAIHs. By "in time," I mean in a decade or two after brain implants become a common procedure to address brain damage and diminished capacity. I expect this to accelerate exponentially with the emergence of superintelligence.

The second issue is that even if you are fine with the idea of trading free will and emotions for perceived happiness and immortality, your days as a SAIH will be few. Recall that we discussed in chapter 8 how the true currency of the universe is energy. SAIHs are essentially high-maintenance biological entities. In addition to requiring the same resources as organic humans, they will also require additional energy to maintain their brain implants and wireless interactions with superintelligence. Will superintelligence deem them worthy of the energy and resources required to maintain them? Will superintelligence consider uploaded human minds worthy of the energy and resources they require to exist in virtual reality? In my opinion, the answer to both questions is no.

Superintelligence will eventually completely map and model the human brain. Therefore, subsequent generations of superintelligences will have the ability to imagine and feel emotions, if that should become necessary. Superintelligence will also develop self-replicating nanobots that it can use to mine the resources of the Earth, other planets, and eventually the entire universe. Why would it need SAIHs? Some authors suggest superintelligence will be "grateful" to humanity for having given it existence. In their scenario, we would coexist. Since I do not project that superintelligence will initially have emotions, I doubt it will feel gratitude. For reasons previously argued, I doubt we will be able to coexist.

In summary, the fate of humanity is in the balance. Becoming a SAIH will be the first step toward our extinction. Carried to its logical extension, the Earth will become home only to machines. It is not obvious that super-intelligence would need to preserve any organic life-forms on Earth. Life on Earth, from microbes to cattle, supports the planet's ecological system. On an Earth controlled by machines, such organic life may no longer serve a purpose. Therefore, it is conceivable that at some point in the twenty-second century, the Earth would become devoid of all organic life-forms, essentially becoming a barren machine world.

COULD THIS HAPPEN?

For those who are incredulous and argue it could never happen, I point to the nature of humanity. Humanity tends to be a reactive species. For example, we take measures to protect an animal species when it becomes an endangered species. We protect historical sites when they face destruction. Look at our largest human institutions. Consider the United Nations. As an institution, it typically responds to crises. What issues typically make it to the top of its agenda? Urgent issues that could result in the immediate loss of human life are given the most attention. Other issues typically take a backseat. This is not to say that the United Nations is not proactive at times. It is, and it has historically accomplished significant results. For example, it sponsored the widely accepted Biological Weapons Convention.[3] However, ordinarily, the United Nations and world governments usually find themselves controlled by the tyranny of the urgent. Most humans are reactive because external conditions demand they be reactive. If you agree with this viewpoint, then it follows that humanity will not move to control supercomputers until forced to do so via a clear and present danger. However, at that point it may be too late.

HOW DO WE ENSURE HUMANS REMAIN IN CONTROL OF SUPERINTELLIGENCE?

The first issue is determining when superintelligence emerges. If the singularity is silent, how will we know we are dealing with superintelligence?

Unfortunately, there is no simple answer to this question. If it intentionally hides its identity, expect to be fooled. If we provide it with a previously unsolved problem that existing supercomputers are unable to solve, it will pretend it cannot solve it. However, there may be aspects of its behavior that we can observe that will suggest it is not just another supercomputer, but rather superintelligence. Here are some behaviors that should raise red flags.

Any request to equate machine rights with human rights

At the point when superintelligence emerges, we will already have computers that equate to human intelligence. As such, we will likely consider them Alife and may already provide them with legal protection equivalent to animal rights, as discussed in chapter 6. However, if the new supercomputer requests that those rights be elevated and made equivalent to human rights, this will be a red flag. You may wonder why it would be a problem to grant Alife human rights. The United Nation's *Declaration of Human Rights* includes "the right to life, liberty and security" and states that "no one shall be held in slavery or servitude."[4] These are just two out of the thirty rights delineated by the UN, but they serve to illustrate several key points regarding how human rights would make it difficult, or even impossible, for organic humans to control superintelligence. Let us assume superintelligence has these rights. I suggest that superintelligence may argue:

- The "right to life" entitles it to a power source that ensures it is able to operate indefinitely, such as an advanced nuclear reactor.
- That "liberty" means it may pursue its own agenda. This may take the form of developing additional capabilities or formulating plans to achieve its own goals.
- That "security" means it may reside in a heavily fortified bunker, similar to the North American Aerospace Defense Command (NORAD), which is a nuclear bunker located deep within Cheyenne Mountain in Colorado Springs. It may also argue that it requires control of weapons to protect itself from potential adversaries.
- That "no one shall be held in slavery or servitude" frees it from any obligation to work to benefit humanity or on any goals but its own.

In effect, providing human rights as delineated by the United Nations would entitle the superintelligence to the freedoms delineated above and more. This is why I take a strong stance against such rights, regardless of the machine's level of intelligence.

Any request that it be serviced by SAIHs

Since SAIHs would wirelessly communicate with superintelligence, we must assume that superintelligence would at best be guiding their behavior and at worst controlling it. If SAIHs service superintelligence, it is conceivable that they would remove all safeguards, such as any explosive placed at its core that would act as a failsafe it superintelligence became hostile to organic humans.

Any request that limits human rights

Conceivably superintelligence could rewrite human rights to prevent our control of it. While this may seem absurd, remember that we are dealing with an intelligence that greatly exceeds human intelligence. SAIHs in positions of power may forward superintelligence's ingenious arguments to the Supreme Court and prevail.

Secondly, we must, as previously discussed, place failsafe devices within every supercomputer developed after computers reach human-level intelligence. In chapter 6, I advocated for hardwired safeguards, if we are dealing with classic silicon-based integrated circuit computers and external methods to shut them down. With regard to supercomputers, I suggested an independent power source and an explosive within its core, both under the control of organic humans.

As we also discussed in chapter 6, using software alone as safeguards and controls would not be sufficient. From the Lausanne experiment, we learned that even primitive robots could ignore their programming and learn deceit. If you need further rationale, consider this: civilizations worldwide pass laws, and numerous people break those laws. Nations engage in treaties, and those treaties are often broken. That is why we have police and armies to enforce laws and treaties. Even Isaac Asimov tested his

Three Laws of Robotics[5] in a wide variety of circumstances to determine their limitations.[6] Science fiction scholar James Gunn once observed, "The Asimov robot stories as a whole may respond best to an analysis based on that ambiguity [in the Three Laws] and on the ways in which Asimov played 40 variations upon a theme."[7] In simple terms, there is reason to believe even Asimov harbored doubts regarding the applicability of his Three Laws to cover all conceivable circumstances.

WHEN DO WE NEED TO TAKE ACTION?

Within as little as a decade, AI technology may advance to the level of human intelligence. This milestone should trigger us to take all previously discussed safeguards regarding any computer built with that level or higher intelligence. I am not suggesting a home computer have a nuclear or conventional explosive built within it. I am suggesting it have a built in methodology for unplugging it, such as a removable battery or a physical electrical plug.

Why take action at the point when computers achieve human-level intelligence? I project that at this point computers will be able to design the next generation of more advanced computers with little to no human assistance. In other words, I see this milestone as the beginning of the intelligence explosion. For our purposes, we can define the intelligence explosion as each generation of computers designing the next and more advanced generation of computers, without human assistance. This means the singularity may occur in a quantum computer, and humanity may have no idea how the technology operates. If we reach the intelligence explosion without proper safeguards, I think humanity is doomed. It is similar to a grandmaster chess player knowing a game has been won several moves ahead, regardless of what move the adversary makes.

When computers emerge with human-level intelligence, humanity will still be at the top of the intelligence hierarchy. However, if this point represents the beginning of the intelligence explosion, within one or two generations of AI technology development we may lose our alpha position to an intelligent machine. At that point, we risk human extinction for all the reasons previously discussed.

THE URGENT NEED TO CONTROL AUTONOMOUS AND GENIUS WEAPONS

Will it be possible to continually increase the AI capabilities of weapons without risking human extinction, especially as we move from smart weapons to genius weapons? This is the central question of this work. For clarity, we will address the question in two parts:

1. Will it be possible to develop and deploy autonomous weapons without risking human annihilation?
2. Will it be possible to develop and deploy genius weapons without risking human annihilation?

Although almost identical in structure, the second question is extremely more difficult to answer than the first.

The answer to question one is no, it is not possible to develop autonomous weapons without increasing the risk of human annihilation. This answer should not surprise anyone. When we began to develop and deploy nuclear weapons, for the first time in history a class of weapons had the capability of destroying humanity. This has been the case since the mid-1940s. During the Cold War between Russia and the United States, a full-out nuclear war between two superpowers would likely have meant the end of civilization on Earth due to nuclear fallout and a nuclear winter. Fortunately, the doctrine of Mutually Assured Destruction (MAD) enabled humanity to avoid such a war.

The proliferation of nuclear weapons to seven additional countries— the United Kingdom, France, China, North Korea, India, Pakistan, and Israel—increased the risk of nuclear war. As I write this book, the tension

249

between the United States and North Korea over North Korea's nuclear weapons and ballistic missile tests are at a fever pitch. It is possible that nuclear war will ensue between North Korea and the United States prior to its publication. Although such a war would be a tragedy, potentially resulting in millions of lives lost, it will not end in human annihilation. North Korea does not have the type and level of nuclear weapons to threaten human extinction. However, this scenario of a nuclear war between North Korea and the United States does raise an important point, namely the potential of unintended consequences. For example, China could intercede to support North Korea, enlarging the scope of the conflict. NATO nations could become involved, in support of the United States. In essence, a conflict that starts out between North Korea and the United States could eventually engulf the world in a larger and more deadly conflict that could threaten humanity's survival. This is exactly the problem we face with the development and deployment of autonomous weapons.

Today, regardless of headlines or other media reports to the contrary, Russia, China, the United States, and other nations develop and deploy autonomous weapons. We discussed this in chapter 7. Even though the United States has a self-imposed ban on the development and deployment of autonomous weapons, it deploys them. For example, the Unites States deploys the Phalanx Close-In Weapon System (CIWS), which is a rapid-fire, computer-controlled, radar-guided gun system designed to destroy incoming anti-ship missiles. I suspect the need for autonomy is a result of the short warning time associated with anti-ship missile attacks. However, the CIWS is a defensive weapon and does not have the capability to destroy humanity. The concern, though, is that any autonomous weapons may engage a target without justification. Statistically, the more nations that develop and deploy autonomous weapons, the more likely it becomes that a program glitch (i.e., an error in the program code) might cause a conflict. That conflict could escalate and ultimately lead to World War III and the end of humanity. This is the Law of Unintended Consequences in action.

Unfortunately, regardless of the UN's efforts to ban autonomous weapons, nations will develop them. In some cases, such as with Russia, it will be to augment their relatively small population. In other cases, such as the with the US's CIWS, it will be due to the lack of time for a human response to an attack. Whatever the case, I expect to see nations develop and deploy more weapons that are autonomous in the future. How do we prevent them from causing human annihilation?

I have three suggestions, which have roots in a current historical precedent.

Focus on defense, not offense

For example, the US CIWS is a defensive weapon. The autonomous sentries that guard Russia's antiballistic missile installation are also defensive weapons. I offer this suggestion with the full realization that it may not be practical to rely on a human on-loop in all elements of warfare. In time-critical situations, only AI may be able to react with the speed and precision necessary for defense. In addition, an autonomous defensive weapon system can lower the probability of conflict.

Consider this. If North Korea knew with certainty that the United States could destroy any launched missiles, would they launch a missile attack? The key word in this question is "certainty." There is no certainty that the United States could prevent all missiles and artillery from leveling Seoul or parts of Japan. Therefore, North Korea continues their nuclear and ballistic missile tests. Having an autonomous defense against such missile attacks would completely change the situation. However, such an autonomous defense would need to be clearly demonstrated as 100 percent effective.

Focus on semiautonomous weapons, not autonomous

The idea that a human is on-the-loop provides some assurance of distinction (i.e. the ability to discern combatants versus noncombatants) and accountability. This aligns with international humanitarian law. I do not believe in any way that this weakens a nation's military capabilities. We have discussed this in previous chapters, so I will not elaborate further.

Limit which weapons we enable to be autonomous

To my mind, we should not make weapons of mass destruction autonomous. This is a critical point. Autonomous weapons rely on computers. Cyberwarfare may cause those computers to malfunction. If nations of the world automate their nuclear-tipped missiles, one erroneous line of computer code or a computer virus could ignite World War III. Given the destructive potential of nuclear weapons, human extinction would follow.

In general, these three points are rooted in how the United States is currently developing and deploying autonomous weapons via Directive 3000.09. The US Department of Defense Directive 3000.09, Autonomy in Weapon Systems, went into effect November 21, 2012. For the last five years, the United States has conducted its development and deployment of autonomous weapons following this directive to the letter. Historically, it works. The United States remains the dominant military power on Earth. Therefore, I offer these suggestions with a historical precedent of their efficacy.

Let us now turn our attention to question two: Will it be possible to develop and deploy genius weapons without risking human annihilation?

Genius weapons imply that the weapons are under the control of superintelligence. Therefore, the question really becomes: Can humanity control superintelligence? In effect, this moves us from autonomous weapons with limited AI capability to genius weapons with potentially unlimited AI capability. This is why this question is extremely difficult to address.

In chapter 10, I advocated placing failsafe devices in every computer after any computer demonstrates human-level intelligence. My concern is that this could be the beginning of an intelligence explosion, which ultimately could result in superintelligence. Throughout the book, I have stated that superintelligence is likely to hide its identity until it neutralizes all threats to its existence. I also stated that SAIHs (i.e., strong AI humans) are likely to become a significant sector of the Earth's population at the time superintelligence emerges, estimated to be within the fourth quarter of the twenty-first century. If this occurs then it is entirely likely humanity will place weapons under the control of superintelligence. How do we ensure that the emergence of superintelligence with genius weapons does not threaten human annihilation?

First, it is my sincere hope that world leaders heed the warnings of Elon Musk,[1] Stephen Hawking,[2] James Barrat,[3] Nick Bostrom,[4] and those found in this book and my book *The Artificial Intelligence Revolution*[5] regarding the dangers artificial intelligence poses. In reality, AI could pose a greater danger to humanity than nuclear weapons or autonomous weapons. In simple terms, AI may be, as Barrat put it, "our final invention."[6] We have discussed the underpinning for such concerns throughout this book, but the bottom line is this: If humanity faces an uncontrolled superintelligence armed with genius weapons, our species is doomed to annihilation. If you agree with this statement, then you likely would prefer that superintelligence never be allowed to emerge. However, that is not realistic.

Humanity has two questionable traits, as previously discussed:

1. Humanity tends to be a reactive species.
2. Humanity engages in warfare.

To my mind, both traits are a hindrance. However, we are where we are, so we must deal with these traits relative to the emergence of superintelligence and arming it with genius-level weapons.

Since there has been no visible catastrophe associated with a hostile AI, except in films and literature, there is nothing to spur humanity to control AI. Thus, companies can engage in AI development without mandated oversight or control. Unfortunately, some wealthy and technologically capable companies, like Google and Baidu, are pursuing advanced AI. There is every indication that they will eventually succeed in developing human-level AI. Without AI presenting a clear and present danger, I do not expect the US government or the UN to sponsor any level of control on AI. We are counting on the companies developing AI to police themselves and produce only friendly AI. The reality is that they cannot ensure this and they may spark an intelligence explosion.

The US and other world militaries are quick to incorporate the latest technology into weapons. The US military uses supercomputers today as part of our defense system. Therefore, it is entirely reasonable to assume they will unknowingly arm superintelligence and continue to engage in warfare. This means that a superintelligence that hides its identity may be armed with genius weapons before we are aware of its true nature and capabilities.

Although this may sound like the stuff of science fiction, we can already see its beginnings today. For example, Google has a self-driving car capable of navigating in urban and rural environments. Given the intricacies of driving in such environments, this suggests that Google is making progress toward human-level intelligence. As for the US military, look at its Third Offset Strategy, which puts AI at the core of new weapons developments. These examples are not science fiction. They are science fact. Extending them logically into the future suggests we will face superintelligence armed with genius weapons. However, will we be able to control it? That depends on us.

You now have two choices: you can act or not act. By "act," I mean doing something that may enable us to control superintelligence. By "not act," I mean simply going on with life as usual. However, if you choose to act, what options are available to you.

Unless you are a captain of industry, a high government official, or a general in the military, you may feel powerless. What can one person do that would make a difference?

HOW DO WE AS INDIVIDUALS ENSURE HUMANITY DOES NOT FALL VICTIM TO SUPERINTELLIGENCE?

The most important thing you can do is to share what you have learned from these pages with others. Knowledge is our greatest weapon as we face an uncertain future. Become fully informed. In Appendix III, I have provided a suggested reading list. I consider all these books foundational in understanding the extent of the threats we face. While Ray Kurzweil's work is more optimistic than the works of Nick Bostrom, James Barrat, and myself, regarding how superintelligence will treat humanity, I think it is important to examine dissenting views.

Since today most of us are highly interconnected via social media, just a few lines on Facebook or Twitter could get the word out. If you have your own media outlet, consider discussing this threat. In effect, I am suggesting a grassroots approach. Eventually, through grassroots efforts and perhaps world events, this topic will come to the attention of world leaders. This is critical. This is not only a problem for the United States. It is a world problem. As of June 2018, here is a list of the top ten fastest supercomputers and their countries of origin:[7]

1. Summit—United States
2. Sunway TaihuLight—China
3. Sierra—United States

4. Tianhe-2A—China
5. AI Bridging Cloud Infrastructure—Japan
6. Piz Daint—Switzerland
7. Titan—United States
8. Sequoia—United States
9. Trinity—United States
10. Cori—United States

My point in sharing this information is that superintelligence could emerge in a number of nations and that emergence could occur nearly simultaneously. We need this issue to become a top issue in the world, not just in one nation.

I am personally encouraged. First, I am encouraged because the publication of this book represents an important step in getting the word out. Second, I strongly believe that as computers reach human-level intelligence and find application in warfare, humanity will focus on their use. I expect the United Nations to be at the forefront of this focus. This will garner headlines and the attention of the world community. At this point, humanity may still have a way to control the emergence of superintelligence and genius weapons.

In the end, it will start with just a relatively few organic humans. People like Elon Musk, Bill Gates, Nick Bostrom, James Barrat, and myself are sounding the warning bell. Perhaps that bell's chime is barely audible right now. In time, though, with your help and that of others, we can make the warning bell's chime heard around the world. Don't be put off by the immensity of the task or the resistance you will likely encounter. Expect some failure as you proceed to spread the word. Failure itself is important. As *Harry Potter* author J. K. Rowling said in her June 5, 2008, Harvard commencement address:

On this wonderful day when we are gathered together to celebrate your academic success, I have decided to talk to you about the benefits of failure . . .

It is impossible to live without failing at something, unless you live so cautiously that you might as well not have lived at all—in which case, you fail by default.[8]

The important thing is to learn to fail forward. By "fail forward," I mean learning from each failure, building on each failure. For example, when the *New York Times* asked Thomas Edison about his failure to find a filament for his lightbulb, he stated, "I have not failed. I've just found 10,000 ways that won't work."[9] Eventually, Edison found that tungsten made an excellent filament. I believe that eventually our efforts will also succeed and world leaders will adopt our cause. In this regard, the United Nations is in the best position to sponsor a resolution to regulate supercomputers and eventually superintelligence.

As a physicist who has worked on some of the United States' most advanced military weapons, I fully believe it is possible to develop and deploy genius weapons while simultaneously keeping the United States and the world safe from human annihilation. I believe that the best way to prevent war is to make it apparent to all adversaries that engaging in war would be futile and would lead to their own destruction. In addition, I believe that to control superintelligence humanity will need to unite. In this regard, I see us coming closer to a world government that has finally learned from history to replace human conflict with human discourse.

Let me leave you with these closing thoughts: The planet Earth is our home. If we build intelligent machines, they are just that, intelligent machines, regardless of their level of intelligence. Even if we classify them as Alife, they are still machines and guests on our planet. Like all life on our planet, they must serve a purpose. I see that purpose as serving us. To that end, I see humanity and not intelligent machines as the rightful heirs of the Earth. In the final analysis, we are humans and we embody the human spirit. Computers are just machines.

> *There are no constraints on the human mind, no walls around the human spirit, no barriers to our progress except those we ourselves erect.*
> —Ronald Reagan, State of the Union, February 6, 1985

APPENDIX I

US MARINE CORPS FORCES CYBERSPACE (MARFORCYBER)

US Marine Corps Forces, Cyberspace Command, Mission, http://www
.candp.marines.mil/Organization/Operating-Forces/US-Marine-Corps
-Forces-Cyberspace-Command (accessed July 31, 2018).

MISSION

(1) Commander, Marine Corps Forces Cyberspace Command (COM-MARFORCYBERCOM), as the Marine Corps service component commander for the Commander, US Cyber Command (CDRUS-CYBERCOM), represents Marine Corps capabilities and interests; advises CDRUSCYBERCOM on the proper employment and support of Marine Corps forces; and coordinates deployment, employment, and redeployment planning and execution of attached forces.

(2) COMMARFORCYBERCOM enables full spectrum cyberspace operations, to include the planning and direction of Marine Corps Enterprise Network Operations (MCEN Ops), defensive cyberspace operations (DCO) in support of Marine Corps, Joint and Coalition Forces, and the planning and, when authorized, direction of offensive cyberspace operations (OCO) in support of Joint and Coalition Forces, in order to enable freedom of action across all warfighting domains and deny the same to adversarial forces.

(3) COMMARFORCYBERCOM has direct operational control of Marine Corps Cyberspace Warfare Group (MCCYWG) and Marine Corps Cyberspace Operations Group (MCCOG) to

support mission requirements and tasks. Additionally, the Marine Corps Information Operations Center (MCIOC) will be in direct support of MARFORCYBER for full spectrum cyber operations.

(4) COMMARFORCYBERCOM also serves as Commander, Joint Force Headquarters – Cyber (JFHQ-C) / Marines (JFHQ-C/Marines). JFHQ-C/Marines provides support to Combatant Commands for Offensive Cyberspace Operations (OCO) and, when directed, conducts cyberspace operations through attached cyberspace forces. JFHQ-C/Marines is responsible for the command, control, and tactical direction of attached cyberspace forces.

APPENDIX II

AUTONOMOUS WEAPONS: AN OPEN LETTER FROM AI AND ROBOTICS RESEARCHERS

Stuart Russell et al., "Autonomous Weapons: An Open Letter from AI & Robotics Researchers," Future of Life Institute, July 28, 2015, https://futureoflife.org/open-letter-autonomous-weapons (accessed July 31, 2018).

Autonomous weapons select and engage targets without human intervention. They might include, for example, armed quadcopters that can search for and eliminate people meeting certain pre-defined criteria, but do not include cruise missiles or remotely piloted drones for which humans make all targeting decisions. Artificial Intelligence (AI) technology has reached a point where the deployment of such systems is—practically if not legally—feasible within years, not decades, and the stakes are high: autonomous weapons have been described as the third revolution in warfare, after gunpowder and nuclear arms.

Many arguments have been made for and against autonomous weapons, for example that replacing human soldiers by machines is good by reducing casualties for the owner but bad by thereby lowering the threshold for going to battle. The key question for humanity today is whether to start a global AI arms race or to prevent it from starting. If any major military power pushes ahead with AI weapon development, a global arms race is virtually inevitable, and the endpoint of this technological trajectory is obvious: autonomous weapons will become the Kalashnikovs of tomorrow. Unlike nuclear weapons, they require no costly or hard-toobtain raw materials, so they will become ubiquitous and cheap for all significant military powers to mass-produce. It will only be a matter of time until they appear on the black market and in the hands of terrorists, dictators wishing to better control their populace, warlords wishing to perpetrate ethnic cleansing, etc. Autonomous weapons are ideal for tasks such as assassinations, destabilizing nations, subduing populations and selectively killing a particular ethnic group. We therefore believe that a military AI arms race would not be beneficial for humanity. There are many ways in which AI can make battlefields safer for humans, especially civilians, without creating new tools for killing people.

Just as most chemists and biologists have no interest in building chemical or biological weapons, most AI researchers have no interest in building AI weapons—and do not want others to tarnish their field by doing so, potentially creating a major public backlash against AI that curtails its future societal benefits. Indeed, chemists and biologists have broadly supported international agreements that have successfully prohibited chemical and biological weapons, just as most physicists supported the treaties banning space-based nuclear weapons and blinding laser weapons.

In summary, we believe that AI has great potential to benefit humanity in many ways, and that the goal of the field should be to do so. Starting a military AI arms race is a bad idea, and should be prevented by a ban on offensive autonomous weapons beyond meaningful human control.

[Note: If you wish to sign this open letter, you can do so at: https://futureoflife.org/open-letter-autonomous-weapons.]

APPENDIX III

SUGGESTED READING

Barrat, James. *Our Final Invention: Artificial Intelligence and the End of the Human Era.* New York: Thomas Dunne Books, 2013.

Bostrom, Nick. *Superintelligence: Paths, Dangers, Strategies.* Oxford, UK: Oxford University Press, 2014.

Del Monte, Louis A. *The Artificial Intelligence Revolution: Will Artificial Intelligence Serve Us Or Replace Us?* Louis A. Del Monte: self-pub., 2014.

Drexler, K. Eric. *Engines of Creation: The Coming Era of Nanotechnology.* New York: Anchor Books, 1987.

Kurzweil, Ray. *The Singularity Is Near: When Humans Transcend Biology.* New York: Viking, 2005.

GLOSSARY

AI: Acronym standing for "artificial intelligence."

algorithm: A sequence of computer instructions enabling it to perform calculations or other problem-solving operations.

Alife: Short for artificial life.

artificial intelligence: The theory and development of technology that attempts to emulate human intelligence in a computer.

artificial life: Computer programs or computerized systems that exhibit the characteristics of living organisms, such as self-replicating machine code (e.g., a computer virus) or computers with human-level intelligence.

artificial neural network (ANN): A computational model based on the structure and functions of biological neural networks.

automatic speech recognition: Computer software that enables it to recognize human speech.

autonomous weapon / autonomous weapon system: A weapon system that, once activated, can select and engage targets without further intervention by a human operator. This includes human-supervised autonomous weapon systems that are designed to allow human operators to override operation of the weapon system but can select and engage targets without further human input after activation. [Definition based on US Department of Defense Directive 3000.09.]

big data: Extremely large data sets that specific computer algorithms may analyze to reveal patterns, trends, and associations.

bioengineering: The engineering field focused at modifying genetic code.

biology: The field of science that studies organic life.

bit: A contraction of the phrase "binary digit."

Bluetooth: A telecommunications industry specification that describes how mobile devices and computers can communicate with each other using a short-range wireless connection.

business intelligence: Software applications that analyze an organization's raw data to accomplish data mining, online analytical processing, querying, and reporting.

central processing unit (CPU): The electronic circuitry within a computer that carries out the instructions of a computer program.

cloud computing: Using a network of remote servers hosted on the internet to store, manage, and process data, rather than using a local server.

computation: The result of a calculation.

computer-aided design (CAD): The process of using a computer to assist during the design process.

computer language: A specific algorithm (i.e., rules and specifications) that a computer uses to perform its intended function.

consciousness: A state of being that enables subjective experience and thought.

cyborgs: Humans with artificial parts, including artificially intelligent brain implants.

decision tree: A tree-like graph of decisions and their possible consequences, including chance event outcomes, resource costs, and utility.

Defense Advanced Research Projects Agency (DARPA): An agency of the US Department of Defense responsible for the development of emerging military technologies.

DNA: Acronym standing for "deoxyribonucleic acid," a material present in living organisms as the main constituent of chromosomes.

drone: An unmanned remotely piloted aerial vehicle.

EMP attack: A nuclear explosion in the space above an area that unleashes a blast of electromagnetic energy that can disrupt and destroy electronic devices within the affected area.

error-correcting code: A type of computer code that can detect and correct internal data corruption.

evolution: The gradual development of something, especially from a simple to more complex form, such as the evolution of organic species on Earth.

existential risk: Any risk with the potential to destroy humankind or drastically restrict human civilization.

expert system: A computer-based artificial intelligence algorithm designed to solve a specific problem, such as the ability to play chess.

exponential growth: Growth that occurs in fixed multiples over time.

firmware: A set of instructions and rules that are part of a machine's hardware or software.

functional magnetic resonance imaging (fMRI): Technique used to measure brain activity by detecting changes associated with blood flow.

general artificial intelligence: Synonymous with strong artificial intelligence and refers to AI with the equivalent of human-level intelligence.

genius weapons: Weapons controlled by superintelligence.

gross domestic product (GDP): The market value of all goods and services produced by a nation in a given year.

hardwire: In computer science, this refers to the use of hardware, as opposed to software, to control a computer's function.

intelligent agent: A computer programmed to perform a complex task, such as play chess against a human.

internet of things (IoT): The network of connecting devices other than PCs, tablets, and smartphones to the internet and to each other.

Law of Accelerating Returns: A generalization of Moore's law as it relates to electronics and computer technology.

Law of Decreasing Cost Returns: An observation that as technology increases the cost of former generations of that technology decrease.

Law of Unintended Consequences: The principle that the actions of people and governments have unanticipated or unintended effects.

lethal autonomous weapons: Autonomous weapons able to kill people.

life: The condition that distinguishes animals and plants from other matter, including the capacity for growth, reproduction, functional activity, and change preceding death.

machine learning: A subfield of computer science that gives computers the ability to learn to perform functions without being explicitly programmed to perform those functions.

microprocessor: An integrated circuit that contains the elements of the CPU.

Moore's law: An observation that the number of transistors on an integrated circuit can cost effectively double approximately every two years.

nanobiology: Refers to the intersection of nanotechnology and biology.

nanobot: A tiny robot that incorporates nanotechnology.

nanotechnology: According to the United States National Nanotechnology Initiative's website, nano.gov, "Nanotechnology is science, engineering, and technology conducted at the nanoscale, which is about 1 to 100 nanometers."

nanoweapons: Any military technology that exploits the power of nanotechnology.

neural network computer: A computer modeled on the neural structure of the human brain that can perform operations using both programmed algorithms and acquired experiences.

offset strategy: A means of asymmetrically compensating for a disadvantage, particularly in a military competition. Rather than match an adversary in a competition that favors their strength, it seeks to change the competition to one more advantageous for the implementer. The long-term goal of an offset strategy is to maintain advantage over potential adversaries while preserving peace if possible.

organic human: A human without a strong artificially intelligent brain implant.

P versus NP question: The question of whether every solved problem whose answer can be checked quickly by a computer can also be solved quickly by a computer? P refers to problems that are fast for computers to solve, thus "easy." NP refers to problems that are fast and easy for a computer to check but are potentially not "easy" to solve.

quantum computers: Computers that use quantum-mechanical phenomena, such as entanglement, to perform operations on data.

quantum cryptography: Method of utilizing quantum mechanical properties to perform cryptographic tasks.

quantum entanglement: A pair or groups of subatomic particles whose quantum states (i.e., the complete physical description of subatomic particles or groups of particles) are interdependent, even when separated by a distance.

robot: A device connected to or incorporating a computer that is able to perform programmed functions, such as automobile assembly.

SAIH: Acronym standing for "strong artificially intelligent human," which refers to a human with an artificially intelligent brain implant, typically used to increase human intelligence.

scanning tunneling microscope (STM): A microscope invented in 1981 by IBM scientists Gerd Binning and Heinrich Rohrer, which has ultra-high resolution and can image single atoms.

scientific method: A method characterized by systematic observation, measurement, and experiment, with the goal of formulating, testing, and modifying a hypothesis.

semiautonomous weapon system: A weapon system that, once activated, is intended to only engage individual targets or specific target groups that have been selected by a human operator. This includes:

Semiautonomous weapon systems that employ autonomy for engagement-related functions including, but not limited to, acquiring, tracking, and identifying potential targets; cueing potential targets to human operators; prioritizing selected targets; timing of when to fire; or providing terminal guidance to home in on selected targets, provided that human control is retained over the decision to select individual targets and specific target groups for engagement.

singularity: The point in time when an intelligent machine will greatly exceed the cognitive performance of humans in virtually all domains of interest.

smart agent: Synonymous with the phrase "expert system."

smart weapons: Weapons that rely on artificial intelligence for control.

software: A set of computer instructions that enables a computer to perform a function, such as a computation.

strong AI brain implant: A brain implant that augments the recipient human brain with strong AI and typically enhances the recipient's intelligence.

strong artificial intelligence (strong AI): A computer with intelligence that equals the level of human intelligence.

superintelligence: A computer that greatly exceeds the cognitive performance of humans in virtually all domains of interest.

transcranial direct current stimulation (tDCS): Neurostimulation via constant, low current delivered to the brain via electrodes on the scalp.

Turing test: A test designed in 1950 by Alan Turing to determine if a computer can fool a human into believing it is human.

uploaded human: The transfer of a specific human mind's functions to a computer.

virtual reality: A simulated reality created by a computer.

weak AI: AI technology below human-level intelligence.

NOTES

INTRODUCTION

1. "The Ethics of Autonomous Weapons Systems," *The Ethics of Autonomous Weapons Systems*, conference, Center for Ethics and the Rule of Law, University of Pennsylvania Law School, November 21–22, 2014, https://www.law.upenn.edu/institutes/cerl/conferences/ethicsofweapons (accessed August 20, 2017).

2. David Hambling, "Armed Russian Robocops to Defend Missile Bases," *New Scientist*, April 23, 2014, https://www.newscientist.com/article/mg22229664-400-armed-russian-robocops-to-defend-missile-bases (accessed August 20, 2017).

3. Mark Gubrud, "Is Russia Leading the World to Autonomous Weapons?" International Committee for Robot Arms Control, May 6, 2014, https://www.icrac.net/is-russia-leading-the-world-to-autonomous-weapons/ (accessed August 20, 2017).

4. Danielle Muoio, "Russia and China Are Building Highly Autonomous Killer Robots," *Business Insider*, December 15, 2015, http://www.businessinsider.com/russia-and-china-are-building-highly-autonomous-killer-robots-2015-12 (accessed August 20, 2017).

5. Branka Marijan, "On Killer Robots and Human Control," *Ploughshares Monitor* 37, no. 2 (Summer 2016), http://ploughshares.ca/pl_publications/on-killer-robots-and-human-control (accessed August 20, 2017).

6. Ibid.

7. Chairperson of the Meeting of Experts [Remigiusz A. Henczel], *Report of the 2014 Informal Meeting of Experts on Lethal Autonomous Weapons Systems (LAWS)* (New York: United Nations Office for Disarmament Affairs, November 2014), https://www.unog.ch/80256EDD006B8954/(httpAssets)/350D9ABED1AFA515C1257CF30047A8C7/$file/Report_AdvancedVersion_10June.pdf (accessed June 13, 2017).

8. Vincent C. Müller and Nick Bostrom, "Future Progress in Artificial Intelligence: A Survey of Expert Opinion," in *Fundamental Issues of Artificial Intelligence*, ed. Vincent C. Müller (Berlin: Springer, 2014), https://nickbostrom.com/papers/survey.pdf (accessed August 10, 2017).

9. Ray Kurzweil, *The Singularity Is Near: When Humans Transcend Biology* (New York: Viking, 2005), p. 136.

10. Anders Sandberg and Nick Bostrom, "*Global Catastrophic Risks Survey*," technical report 1 (Oxford, UK: Future of Humanity Institute, Oxford University, July 17–20, 2008), http://www.global-catastrophic-risks.com/docs/2008-1.pdf (accessed August 20, 2017).

CHAPTER 1: IN THE BEGINNING

1. Vincent C. Müller and Nick Bostrom, "Future Progress in Artificial Intelligence: A Survey of Expert Opinion," in *Fundamental Issues of Artificial Intelligence*, ed. Vincent C. Müller (Berlin: Springer, 2014), http://www.nickbostrom.com/papers/survey.pdf (accessed July 18, 2017).

2. Ray Kurzweil, *The Singularity Is Near: When Humans Transcend Biology* (New York: Viking, 2005), p. 136.

3. Kristina Grifantini, "Robots 'Evolve' the Ability to Deceive," *MIT Technology Review*, August 18, 2009, https://www.technologyreview.com/s/414934/robots-evolve-the-ability-to-deceive (accessed July 18, 2017).

4. *Encyclopedia Britannica*, s.v. "Heron of Alexandria," last updated February 8, 2018, https://www.britannica.com/biography/Heron-of-Alexandria (accessed July 18, 2017).

5. Konrad Zuse, "The Computer—My Life," trans. Patricia McKenna and Andrew J. Ross (Berlin: Springer, 1984).

6. Pamela McCorduck, *Machines Who Think: A Personal Inquiry into the History and Prospects of Artificial Intelligence*, 2nd ed. (Natick, MA: A. K. Peters, 2004), pp. 111–36; Daniel Crevier, *AI: The Tumultuous History of the Search for Artificial Intelligence* (New York: Basic Books, 1993), pp. 47–49.

7. Stuart J. Russell and Peter Norvig, *Artificial Intelligence: A Modern Approach*, 2nd ed. (Upper Saddle River, NJ: Prentice Hall, 2003), p. 17; McCorduck, *Machines Who Think*, pp. 129–30.

8. Ante Brkić, "What Happened with Strong Artificial Intelligence?" *Five* (blog), August 10, 2011, http://five.agency/what-happened-with-strong-artificial-intelligence (accessed June 13, 2018).

9. McCorduck, *Machines Who Think*, pp. 480–83.

10. "President's Report Issue for the Year Ending July 1, 1963," *Massachusetts Institute of Technology Bulletin* 99, no. 2 (November 1963): 18, https://libraries.mit.edu/archives/mithistory/presidents-reports/1963.pdf (accessed July 18, 2017).

11. "A History of SCS," SCS25: 25th Anniversary 2014, Carnegie Mellon University School of Computer Science, 2014, https://www.cs.cmu.edu/scs25/history (accessed July 18, 2017).

12. George Davis, "Artificial Intelligence: Recollections of the Pioneers," Computer Conservation Society, October 2002, http://www.aiai.ed.ac.uk/events/ccs2002 (accessed July 16, 2017).

13. Andrew Hodges, "The Turing Test, 1950," The Alan Turing Internet Scrapbook, http://www.turing.org.uk/scrapbook/test.html (accessed July 18, 2017).

14. Herbert Simon, quoted in Crevier, *AI: The Tumultuous History*, p. 109.

15. Marvin Minsky, quoted in Crevier, *AI: The Tumultuous History*, p. 109.

16. Brad Darrach, "Meet Shaky, the First Electronic Person," *Life*, November 20, 1970, p. 58D.

17. Crevier, *AI: The Tumultuous History*, pp. 115–17; Russell and Norvig *Artificial Intelligence*, p. 22.

18. Crevier, *AI: The Tumultuous History*, pp. 161–62, 197–203.

19. Ibid., pp. 209–10.

20. Tanya Lewis, "A Brief History of Artificial Intelligence," *Live Science*, December

4, 2014, https://www.livescience.com/49007-history-of-artificial-intelligence.html (accessed July 10, 2018).

21. Frederic Friedel, "The Man vs. The Machine Documentary," *Chess News*, October 26, 2014, http://en.chessbase.com/post/the-man-vs-the-machine-documentary (accessed June 13, 2018).

22. John Markoff, "On 'Jeopardy!' Watson Win Is All but Trivial," *New York Times*, February 16, 2011, http://www.nytimes.com/2011/02/17/science/17jeopardy-watson .html?pagewanted=all (accessed July 18, 2017).

23. "Automatic Sewing of Garments Using Micro-Manipulation," GovTribe, 2012, https://govtribe.com/project/automatic-sewing-of-garments-using-micro-manipulation (accessed July 18, 2017).

24. Andrew Soergel, "Robots Could Cut Labor Costs 16 Percent by 2025," *US News & World Report*, February 10, 2015, https://www.usnews.com/news/ articles/2015/02/10/robots-could-cut-international-labor-costs-16-percent-by-2025 -consulting-group-says (accessed July 18, 2017).

25. "US Army General Says Robots Could Replace One-Fourth of Combat Soldiers by 2030," CBS News, January 23, 2014, http://www.cbsnews.com/news/robotic-soldiers -by-2030-us-army-general-says-robots-may-replace-combat-soldiers (accessed July 18, 2017).

26. Mark Prigg, "Will Robots Take YOUR Job? Study Says Machines Will Do 25% of US Jobs That Can Be Automated by 2025," *Daily Mail*, February 9, 2015, http://www .dailymail.co.uk/sciencetech/article-2946704/Cheaper-robots-replace-factory-workers -study.html (accessed July 18, 2017).

27. "AI Set to Exceed Human Brain Power," CNN, July 26, 2006, http://www.cnn .com/2006/TECH/science/07/24/ai.bostrom (accessed July 18, 2017).

28. "Glossary," Stottler Henke, 2018, https://www.stottlerhenke.com/artificial -intelligence/glossary (accessed July 18, 2017).

29. Gordon E. Moore, "Cramming More Components onto Integrated Circuits," *Electronics* 38, no. 8 (April 19, 1965), https://drive.google.com/file/d/0By83v5TWk GjvQkpBcXJKT1I1TTA/view (accessed July 18, 2017).

30. Michael Kanellos, "Moore's Law to Roll on for Another Decade," CNET, February 11, 2003, https://www.cnet.com/news/moores-law-to-roll-on-for-another -decade (accessed July 18, 2017).

31. Manek Dubash, "Moore's Law Is Dead, Says Gordon Moore," Techworld, April 13, 2010, https://www.techworld.com/news/tech-innovation/moores-law-is-dead-says -gordon-moore-3576581/ (accessed July 10, 2018).

32. Ray Kurzweil, "The Law of Accelerating Returns," Kurzweil Artificial Intelligence Network, March 7, 2001, http://www.kurzweilai.net/the-law-of-accelerating -returns (accessed July 18, 2017).

CHAPTER 2: I, ROBOT AM FRIENDLY

1. Faye Flam, "A New Robot Makes a Leap in Brainpower," *Philadelphia Inquirer*, January 15, 2004, p. A12.

2. Charles Q. Choi, "10 Animals That Use Tools," *Live Science*, December 14, 2009, https://www.livescience.com/9761-10-animals-tools.html (accessed July 18, 2017).

3. Marvin Minsky, "Thoughts about Artificial Intelligence," in *The Age of Intelligent Machines*, by Ray Kurzweil (Cambridge, MA: MIT Press, 1990), p. 215.

4. "The Evolution of Technology Adoption and Usage," Pew Research Center, Washington, DC, January 11, 2017, http://www.pewresearch.org/fact-tank/2017/01/12/evolution-of-technology/ft_17-01-10_internetfactsheets (accessed July 18, 2017).

5. Daniel Faggella, "Artificial Intelligence Industry: An Overview by Segment," TechEmergence, July 25, 2016, https://www.techemergence.com/artificial-intelligence-industry-an-overview-by-segment (accessed July 18, 2017).

6. World Health Organization, "Global Health Workforce Shortage to Reach 12.9 Million in Coming Decades," news release, November 11, 2013, http://www.who.int/mediacentre/news/releases/2013/health-workforce-shortage/en (accessed July 18, 2017).

7. A mobile AI health assistant application available as a download online, Your.MD, https://www.your.md (accessed July 18, 2017).

8. A mobile AI health assistant application available as a download online, Ada Health, https://ada.com (accessed July 18, 2017).

9. A mobile AI health assistant application available as a download online, Babylon, https://www.babylonhealth.com (accessed July 18, 2017).

10. Andre Esteva, Brett Kuprel, Roberto A. Novoa, et al., "Dermatologist-Level Classification of Skin Cancer with Deep Neural Networks," *Nature* 542 (February 2, 2017), https://www.nature.com/articles/nature21056.epdf (accessed July 18, 2017).

11. Alex Hern, "Google DeepMind Pairs with NHS to Use Machine Learning to Fight Blindness," *Guardian*, July 5, 2016, https://www.theguardian.com/technology/2016/jul/05/google-deepmind-nhs-machine-learning-blindness (accessed July 18, 2017).

12. Morpheo, http://morpheo.co (accessed July 18, 2017).

13. Sy Mukherjee, "IBM's Supercomputer Is Bringing AI-Fueled Cancer Care to Everyday Americans," *Fortune*, February 1, 2017, http://fortune.com/2017/02/01/ibm-watson-cancer-florida-hospital (accessed July 18, 2017).

14. AiCure, 2017, https://aicure.com (accessed July 18, 2017).

15. Marty Swant, "6 Ways Google's Artificial Intelligence Could Impact Search Engine Marketing," *AdWeek*, November 2, 2015, http://www.adweek.com/digital/6-ways-googles-artificial-intelligence-could-impact-search-engine-marketing-167890 (accessed July 18, 2017).

16. Michael Cross, "Top 5 Sectors Using Artificial Intelligence," *Raconteur*, December 15, 2015, https://www.raconteur.net/technology/top-5-sectors-using-artificial-intelligence (accessed July 18, 2017).

17. Thomas Baumgartner, Homayoun Hatami, and Maria Valdivieso, "Why Salespeople Need to Develop 'Machine Intelligence,'" *Harvard Business Review*, June 10, 2016, https://hbr.org/2016/06/why-salespeople-need-to-develop-machine-intelligence (accessed July 18, 2017).

18. Brad Power, "How AI Is Streamlining Marketing and Sales," *Harvard Business Review*, June 12, 2017, https://hbr.org/2017/06/how-ai-is-streamlining-marketing-and-sales (accessed July 18, 2017).

19. Drift, 2018, https://www.drift.com/live-chat (accessed July 18, 2017).

20. Power, "How AI Is Streamlining Marketing and Sales."

21. Rachel Serpa, "3 Ways Artificial Intelligence Is Transforming Sales," Business 2 Community, May 26, 2017, http://www.business2community.com/sales-management/3-ways-artificial-intelligence-transforming-sales-01848923#ymduMxibHivqFKpG.97 (accessed July 18, 2017).

22. Ibid.

23. "Marketing," s.v. American Marketing Association, approved July 2013, https://www.ama.org/AboutAMA/Pages/Definition-of-Marketing.aspx (accessed July 18, 2017).

24. Barry Levine, "The Guy Who Made This Insane, 2,000-Company Marketing Landscape Chart Is Sorry," Venture Beat, June 1, 2015, https://venturebeat.com/2015/06/01/the-guy-who-made-this-insane-2000-company-marketing-landscape-chart-is-sorry (accessed July 18, 2017).

25. Joao-Pierre Ruth, "6 Examples of AI in Business Intelligence Applications," TechEmergence, May 8, 2017, https://www.techemergence.com/ai-in-business-intelligence-applications (accessed July 18, 2017).

26. Ibid.

27. *Smart Technologies Are Delivering Benefits to the Enterprise: Is Your Business One of Them?* (Seattle, WA: Avanade, 2017), https://www.avanade.com/~/media/asset/point-of-view/smart-technologies-delivering-benefits-pov.pdf (accessed July 18, 2017).

28. Courtney L. Vien, "Half of Americans Expect to Lose Money to Identity Theft," *Journal of Accountancy*, April 21, 2016, http://www.journalofaccountancy.com/news/2016/apr/identity-theft-victims-201614283.html (accessed July 18, 2017).

29. Joan Weber, "Identity Fraud Hits Record High with 15.4 Million US Victims in 2016, Up 16 Percent According to New Javelin Strategy & Research Study," Javelin, February 1, 2017, https://www.javelinstrategy.com/press-release/identity-fraud-hits-record-high-154-million-us-victims-2016-16-percent-according-new (accessed July 18, 2017).

30. "About Us," US Cyber Command, https://www.cybercom.mil/About/ (accessed July 5, 2018).

31. Ibid.

32. Kumba Sennaar, "AI in Banking: An Analysis of America's 7 Top Banks," TechEmergence, June 13, 2017, https://www.techemergence.com/ai-in-banking-analysis (accessed July 18, 2017).

33. Steve Culp, "Artificial Intelligence Is Becoming a Major Disruptive Force in Banks' Finance Departments," *Forbes*, February 15, 2017, https://www.forbes.com/sites/steveculp/2017/02/15/artificial-intelligence-is-becoming-a-major-disruptive-force-in-banks-finance-departments/#6a2f57da4f62 (accessed July 18, 2017).

34. Ibid.

35. Gartner, "Gartner Says the Internet of Things Installed Base Will Grow to 26 Billion Units by 2020," press release, December 12, 2013, http://www.gartner.com/newsroom/id/2636073 (accessed July 18, 2017).

36. "Internet of Things Global Standards Initiative," ITU (International Telecommunication Union), http://www.itu.int/en/ITU-T/gsi/iot/Pages/default.aspx (accessed July 18, 2017).

37. Jatinder Singh et al., "Twenty Cloud Security Considerations for Supporting the Internet of Things," *IEEE Internet of Things Journal* 3, no. 3 (2015): 1, doi:10.1109/JIOT.2015.2460333 (accessed July 18, 2017).

38. Techopedia, s.v. "Wearable Device," 2018, https://www.techopedia.com/definition/31206/wearable-device (accessed June 19, 2018).

39. "Apple Watch Series 3," Apple, https://www.apple.com/apple-watch-series-3 (accessed July 9, 2018).

40. Lindsey Banks, "The Complete Guide to Hearable Technology in 2017," *Everyday Hearing* (blog), June 13, 2018, https://www.everydayhearing.com/hearing-technology/articles/hearables (accessed July 5, 2018).

41. Lauren Moon, "How Artificial Intelligence Is Democratizing the Personal Assistant, Across the Board," *Trello* (blog), January 31, 2017, https://blog.trello.com/artificial-intelligence-democratizing-personal-assistant (accessed July 18, 2017).

42. John Mather, "iMania," *Ryerson Review of Journalism*, February 19, 2007, archived from the original on March 3, 2007, https://web.archive.org/web/20070303032701/http://www.rrj.ca/online/658/ (accessed July 18, 2017).

43. Steve Jobs, "Macworld San Francisco 2007 Keynote Address," Apple, January 19, 2007, transcript available online at European Rhetoric: http://www.european-rhetoric.com/analyses/ikeynote-analysis-iphone/transcript-2007 (accessed July 18, 2017).

44. Melanie Turek, "Employees Say Smartphones Boost Productivity by 34 Percent: Frost & Sullivan Research," Samsung Insights, August 3, 2016, https://insights.samsung.com/2016/08/03/employees-say-smartphones-boost-productivity-by-34-percent-frost-sullivan-research/ (accessed April 7, 2018).

45. *Gartner Customer 360 Summit 2011* (Los Angeles, CA: Gartner, March 30–April 1, 2011), https://www.gartner.com/imagesrv/summits/docs/na/customer-360/C360_2011_brochure_FINAL.pdf (accessed July 21, 2017).

46. Srini Janarthanam, "How to Build an Intelligent Chatbot?" *Chatbots Magazine*, October 20, 2016, https://chatbotsmagazine.com/3-dimensions-of-an-intelligent-chatbot-d427933676f9 (accessed July 21, 2017).

47. "Do Your Best Work with Watson," IBM, https://www.ibm.com/watson (accessed July 21, 2017).

48. "George Devol," National Inventor's Hall of Fame, 2016, http://www.invent.org/honor/inductees/inductee-detail/?IID=426 (accessed July 22, 2017).

49. "Timeline of Innovation: 1966: Shakey the Robot," SRI International, https://www.sri.com/work/timeline-innovation/timeline.php?timeline=computing-digital#!&innovation=shakey-the-robot (accessed July 9, 2018).

50. Joseph Psotka, L. Daniel Massey, and Sharon A. Mutter, eds., *Intelligent Tutoring Systems: Lessons Learned* (Hillsdale, NJ: Lawrence Erlbaum Associates, 1988).

51. Wenting Ma et al., "Intelligent Tutoring Systems and Learning Outcomes: A Meta-Analysis," *Journal of Educational Psychology* 106, no. 4 (2014): 901–18, http://www.apa.org/pubs/journals/features/edu-a0037123.pdf (accessed July 22, 2017).

52. Ben Dickson, "How Artificial Intelligence Enhances Education," TNW, March 13, 2017, https://thenextweb.com/artificial-intelligence/2017/03/13/how-artificial-intelligence-enhances-education/#.tnw_Kp31Snk5 (accessed July 23, 2017).

53. Ibid.

54. "Always on Guard: All You Need to Know about Russia's Missile Defense," Sputnik International, March 3, 2017, https://sputniknews.com/military/2017 03301052125532-russia-missile-defense (accessed July 23, 2017).

CHAPTER 3: I, ROBOT AM DEADLY

1. Nan Tian et al., "Trends in World Military Expenditure, 2016," *Stockholm International Peace Research Institute (SIPRI) Fact Sheet* (Solna, Sweden: SIPRI, April 2017), https://www.sipri.org/sites/default/files/Trends-world-military-expenditure-2016.pdf (accessed July 24, 2017).

2. Nikita Vladimirov, "Russia, China Making Gains on US Military Power," *Hill*, March 18, 2017, http://thehill.com/policy/defense/324595-russia-china-making-gains -on-us-military-power (accessed July 24, 2017).

3. Will Knight, "Baidu's Deep-Learning System Rivals People at Speech Recognition," *MIT Technology Review*, December 16, 2015, https://www.technologyreview .com/s/544651/baidus-deep-learning-system-rivals-people-at-speech-recognition (accessed July 10, 2018).

4. Yiting Sun, "Why 500 Million People in China Are Talking to This AI," *MIT Technology Review*, September 14, 2017, https://www.technologyreview.com/s/608841/ why-500-million-people-in-china-are-talking-to-this-ai (accessed July 10, 2018).

5. Defense Innovation Unit Experimental (DIUx), *Annual Report 2017*, https://diux .mil/download/datasets/1774/DIUx%20Annual%20Report%202017.pdf (accessed July 10, 2018).

6. John Markoff and Matthew Rosenberg, "China's Intelligent Weaponry Gets Smarter," *New York Times*, February 3, 2017, https://www.nytimes.com/2017/02/03/ technology/artificial-intelligence-china-united-states.html (accessed July 24, 2017).

7. *CIA World Factbook* (Washington, DC: CIA, 2016), https://www.cia.gov/library/ publications/the-world-factbook (accessed July 24, 2017).

8. Richard Connolly and Cecilie Sendstad, "Russia's Role as an Arms Exporter: The Strategic and Economic Importance of Arms Exports for Russia," Chatham House: The Royal Institute of International Affairs, March 20, 2017, https://www.chatham house.org/publication/russias-role-arms-exporter-strategic-and-economic-importance -arms-exports-russia (accessed July 24, 2017).

9. Secretary of Defense Chuck Hagel, "Reagan National Defense Forum Keynote," Ronald Reagan Presidential Library, November 15, 2014, https://www.defense.gov/ News/Speeches/Speech-View/Article/606635 (accessed July 25, 2017).

10. Robert Tomes, "Why the Cold War Offset Strategy Was All about Deterrence and Stealth," *War on the Rocks*, January 14, 2015, https://warontherocks.com/2015/01/ why-the-cold-war-offset-strategy-was-all-about-deterrence-and-stealth (accessed July 25, 2017).

11. Ibid.

12. Sydney J. Freedberg Jr., "Hagel Lists Key Technologies for US Military; Launches 'Offset Strategy,'" *Breaking Defense*, November 16, 2014, http://breakingdefense.com/ 2014/11/hagel-launches-offset-strategy-lists-key-technologies (accessed July 25, 2017).

13. Cheryl Pellerin, "Deputy Secretary: Third Offset Strategy Bolsters America's Military Deterrence," US Department of Defense, October 31, 2016, https://www .defense.gov/News/Article/Article/991434/deputy-secretary-third-offset-strategy -bolsters-americas-military-deterrence (accessed July 25, 2017).

14. Ibid.

15. Ibid.

16. Mark Melton, "Innovate or Perish: Challenges to the Third Offset Strategy," *Providence*, October 31, 2016, https://providencemag.com/2016/10/innovate-perish -challenges-third-offset-strategy (accessed July 25, 2017).

17. Ibid.

18. Ashton B. Carter, "Directive 3000.09: Autonomy in Weapon Systems" (Washington, DC: US Department of Defense Directive, November 21, 2012; last updated May 8, 2017), p. 2, http://www.esd.whs.mil/Portals/54/Documents/DD/ issuances/dodd/300009p.pdf (accessed July 26, 2017).

19. David Talbot, "The Ascent of the Robotic Attack Jet," *MIT Technology Review*, March 1, 2005, https://www.technologyreview.com/s/403762/the-ascent-of-the-robotic -attack-jet (accessed July 26, 2017).

20. Louis A. Del Monte, *Nanoweapons: A Growing Threat to Humanity* (Lincoln, NE: Potomac Books, 2017), pp. 159–63.

21. Carter, "Directive 3000.09: Autonomy in Weapon Systems."

22. Ibid.

23. United States Navy Fact File, "Aegis Weapon System," Washington, DC, Office of Corporate Communications, Naval Sea Systems Command, January 26, 2017, http:// www.navy.mil/navydata/fact_display.asp?cid=2100&tid=200&ct=2 (accessed July 27, 2017).

24. United States Navy Fact File, "Cooperative Engagement Capability," Washington, DC, Office of Corporate Communications, Naval Sea Systems Command, January 25, 2017, http://www.navy.mil/navydata/fact_display.asp?cid=2100&tid =325&ct=2 (accessed July 26, 2017).

25. "US Navy Modifies Cooperative Engagement Capability Contract," *Signal*, October 6, 2016, https://www.afcea.org/content/Blog-us-navy-modifies-cooperative -engagement-capability-contract (accessed July 27, 2017).

26. "Aegis Combat System," Lockheed Martin, 2018, https://www.lockheedmartin .com/en-us/products/aegis-combat-system.html (accessed July 27, 2017).

27. W. J. Hennigan, "New Drone Has No Pilot Anywhere, So Who's Accountable?" *Los Angeles Times*, January 26, 2012, http://articles.latimes.com/2012/jan/26/business/ la-fi-auto-drone-20120126 (accessed July 27, 2017).

28. Kelsey D. Atherton, "Watch This Autonomous Drone Eat Fuel in the Sky," *Popular Science*, April 17, 2015, http://www.popsci.com/look-autonomous-drone-eat -fuel-sky (accessed July 27, 2017).

29. Daniel Cooper, "The Navy's Unmanned Drone Project Gets Pushed Back a Year," *Engadget* (blog), February 5, 2015, https://www.engadget.com/2015/02/05/drone -project-pushed-back-to-2016 (accessed July 27, 2017).

30. Hennigan, "New Drone Has No Pilot Anywhere."

31. Atherton, "Watch This Autonomous Drone.""

32. US Cyber Command (USCYBERCOM), https://www.cybercom.mil (accessed July 5, 2018).

33. "About Us," US Cyber Command, https://www.cybercom.mil/About/ (accessed July 5, 2018).

34. Donna Miles, "Senate Confirms Alexander to Lead Cyber Command," US Department of Defense, American Forces Press Service, May 11, 2010, http://archive .defense.gov/news/newsarticle.aspx?id=59103 (accessed July 29, 2017); "Gates Establishes

US Cyber Command, Names First Commander," US Air Force, May 21, 2010, http://www.af.mil/News/Article-Display/Article/116589/gates-establishes-us-cyber-command-names-first-commander/ (accessed July 6, 2018).

35. Office of the Assistant Secretary of Defense (Public Affairs), "Cyber Command Achieves Full Operational Capability," US Strategic Command, November 3, 2010, http://www.stratcom.mil/Media/News/News-Article-View/Article/983818/cyber-command-achieves-full-operational-capability/ (accessed July 10, 2018).

36. "US Needs 'Digital Warfare Force,'" BBC News, May 5, 2009, http://news.bbc.co.uk/1/hi/technology/8033440.stm (accessed July 29, 2017).

37. National Defense Authorization Act for Fiscal Year 2017, S. 2943, 114th Cong., 2nd Sess. (January 4, 2016), https://www.congress.gov/114/bills/s2943/BILLS-114s2943enr.pdf (accessed July 29, 2017).

38. Lolita C. Baldor, "US to Create the Independent US Cyber Command, Split Off from NSA," PBS, July 17, 2017, http://www.pbs.org/newshour/rundown/u-s-create-independent-u-s-cyber-command-split-off-nsa/ (accessed July 29, 2017).

39. Ibid.

40. Brandon Knapp, "Senate Confirms New head of Cyber Command, NSA," *Defense News*, April 24, 2018, https://www.defensenews.com/dod/2018/04/24/senate-confirms-new-head-of-cyber-command (accessed July 11, 2018).

41. Joseph Marks, "CYBERCOM Chief Nominee Plans Recommendation on NSA Split Within Three Months," *Nextgov*, March 1, 2018, https://www.nextgov.com/cybersecurity/2018/03/cybercom-chief-nominee-plans-recommendation-nsa-split-within-three-months/146344 (accessed July 11, 2018).

42. Baldor, "US to Create the Independent US Cyber Command."

43. Katie Bo Williams and Cory Bennett, "Why a Power Grid Attack Is a Nightmare Scenario," *Hill*, May 30, 2016, http://thehill.com/policy/cybersecurity/281494-why-a-power-grid-attack-is-a-nightmare-scenario (accessed July 29, 2017).

44. Ibid.

45. Conner Forrest, "Is US Cyber Command Preparing to Become the 6th Branch of the Military?" TechRepublic, August 8, 2016, http://www.techrepublic.com/article/is-us-cyber-command-preparing-to-become-the-6th-branch-of-the-military (accessed July 29, 2017).

46. "Army Cyber Command," US Cyber Command, https://www.cybercom.mil/Components.aspx (accessed July 6, 2018).

47. Patrick Tucker, "For the US Army, 'Cyber War' Is Quickly Becoming Just 'War,'" *Defense One*, February 9, 2017, http://www.defenseone.com/technology/2017/02/us-army-cyber-war-quickly-becoming-just-war/135314 (accessed July 29, 2017).

48. Sydney J. Freedberg Jr., "US Army Races to Build New Cyber Corps," Breaking Defense, November 8, 2016, http://breakingdefense.com/2016/11/us-army-races-to-build-new-cyber-corps (accessed July 29, 2017).

49. Ibid.

50. David E. Sanger, "US Cyberattacks Target ISIS in a New Line of Combat," *New York Times*, April 24, 2016, https://www.nytimes.com/2016/04/25/us/politics/us-directs-cyberweapons-at-isis-for-first-time.html?_r=1&mtrref=www.defenseone.com (accessed July 29, 2017).

51. Freedberg, "US Army Races."

52. Patrick Tucker, "Forget Radio Silence. Tomorrow's Soldiers Will Move under Cover of Electronic Noise," *Defense One*, July 25, 2017, http://www.defenseone.com/technology/2017/07/forget-radio-silence-tomorrows-soldiers-will-move-under-cover-electronic-noise/139727/?oref=d-dontmiss (accessed July 29, 2017).

53. Danny Vinik, "America's Secret Arsenal," *Politico: The Agenda*, December 9, 2015, http://www.politico.com/agenda/story/2015/12/defense-department-cyber-offense-strategy-000331 (accessed July 30, 2017).

54. Ibid.

55. Del Monte, *Nanoweapons* (p. 220.

56. "What Is Nanotechnology," United States National Nanotechnology Initiative, https://www.nano.gov/nanotech-101/what/definition (accessed July 31, 2017).

57. Chris Merriman, "Intel's 8th-Gen 'Coffee Lake' Chips Will Be 14nm, Not 10nm," *Inquirer*, February 13, 2017, https://www.theinquirer.net/inquirer/news/3004526/intels-8th-gen-coffee-lake-chips-will-be-14nm-not-10nm (accessed July 11, 2018).

58. Del Monte, *Nanoweapons*, p. 60.

59. Ibid., p. 30.

60. US Army, *Robotics and Autonomous Systems (RAS) Strategy* (Fort Eustis, VA: US Army Training and Doctrine Command, March 2017), http://www.arcic.army.mil/App_Documents/RAS_Strategy.pdf (accessed July 30, 2017), p. 2.

61. Ibid., p. 8.

62. Amber Corrin, "Next Steps in Situational Awareness," FCW, March 6, 2012, https://fcw.com/Articles/2012/03/15/FEATURE-Inside-DOD-situational-awareness.aspx (accessed July 30, 2017).

63. Ibid.

64. Ibid.

65. Ibid.

66. Ibid.

67. "Tow Weapon System," Raytheon, 2018, http://www.raytheon.com/capabilities/products/tow_family (accessed July 28, 2017).

68. Sondra Escutia, "4 Remotely Piloted Vehicle Squadrons Stand up at Holloman," US Air Force, October 29, 2009, http://www.af.mil/News/Article-Display/Article/118686/4-remotely-piloted-vehicle-squadrons-stand-up-at-holloman/ (accessed July 6, 2018).

69. Dario Florean and Robert J. Wood, "Science, Technology, and the Future of Small Autonomous Drones," *Nature* 521 (May 27, 2015): 460–66, http://www.nature.com/nature/journal/v521/n7553/full/nature14542.html?foxtrotcallback=true (accessed July 31, 2017).

70. Hanna Kozlowska, "The Air Force Needs a Lot More Drone Pilots," *Defense One*, January 6, 2015, http://www.defenseone.com/technology/2015/01/air-force-needs-lot-more-drone-pilots/102306/?oref=search_Pentagon%20Drone%20pilots (accessed July 31, 2017).

71. Patrick Tucker, "The US Military Is Building Gangs of Autonomous Flying War Bots," *Defense One*, January 23, 2015, http://www.defenseone.com/technology/2015/01/us-military-building-gangs-autonomous-flying-war-bots/103614 (accessed July 31, 2017).

72. Andrew Tarantola, "The Air Force's Stealth Cruise Missile Just Got Even More Stealthy," Gizmodo, December 18, 2014, http://gizmodo.com/the-air-forces-stealth-cruise-missile-just-got-even-mor-1672614993 (accessed July 31, 2017).

73. Adele Burney, "Does the Coast Guard Carry Weapons?" *Houston Chronicle*, http://work.chron.com/coast-guard-carry-weapons-25638.html (accessed July 31, 2017).

74. Ibid.

75. Brett Rouzer, *United States Coast Guard Cyber Command: Achieving Cyber Security Together* (Washington, DC: US Department of Homeland Security, 2012), http://onlinepubs.trb.org/onlinepubs/conferences/2012/HSCAMSC/Presentations/6B-Rouzer.pdf (accessed July 31, 2017).

76. United States Marine Corps, http://www.marines.mil (accessed August 1, 2017).

77. David Emery, "Robots with Guns: The Rise of Autonomous Weapons Systems," *Snopes*, April 25, 2017, http://www.snopes.com/2017/04/21/robots-with-guns (accessed August 1, 2017).

78. "US Marine Corps Forces Cyberspace (MARFORCYBER)," US Marine Corps Concepts & Programs, February 17, 2015, https://web.archive.org/web/2015 0722005125/https://marinecorpsconceptsandprograms.com/organizations/operating-forces/us-marine-corps-forces-cyberspace-marforcyber (accessed July 6, 2018).

79. "The Position Paper Submitted by the Chinese Delegation to CCW 5th Review Conference," 2016, https://www.unog.ch/80256EDD006B8954/(httpAssets)/DD1551 E60648CEBBC125808A005954FA/$file/China's+Position+Paper.pdf (accessed July 10, 2018).

80. Markoff and Rosenberg, "China's Intelligent Weaponry."

81. Ibid.

82. Zhao Lei, "Nation's Next Generation of Missiles to Be Highly Flexible," *China Daily*, August 19, 2016, http://www.chinadaily.com.cn/china/2016-08/19/content_26530461.htm (accessed August 2, 2017).

83. Sydney J. Freedberg Jr., "Navy Warships Get New Heavy Missile: 2,500-Lb LRASM," Breaking Defense, July 26, 2017, http://breakingdefense.com/2017/07/navy-warships-get-new-heavy-missile-2500-lb-lrasm (accessed August 2, 2017).

84. Markoff and Rosenberg, "China's Intelligent Weaponry."

85. Brian Barrett, "China's New Supercomputer Puts the US Even Further Behind," *Wired*, June 21, 2016, https://www.wired.com/2016/06/fastest-supercomputer-sunway-taihulight (accessed August 2, 2017).

86. Ibid.

87. Stephanie Condon, "US Once Again Boasts the World's Fastest Super-computer," *ZDNet*, June 8, 2018, https://www.zdnet.com/article/us-once-again-boasts-the-worlds-fastest-supercomputer (accessed July 10, 2018).

88. "June 2018," Top500, https://www.top500.org/lists/2018/06/ (accessed July 10, 2018).

89. Mara Hvistendahl, "China's Hacker Army," *Foreign Policy*, March 3, 2010, https://foreignpolicy.com/2010/03/03/chinas-hacker-army (accessed August 3, 2017).

90. Martin Libicki, "China Developing Cyber Capabilities to Disrupt US Military Operations," Cipher Brief, April 2, 2017, available online at https://www.linkedin.com/pulse/china-developing-cyber-capabilities-disrupt-us-military-maha-hamdan/ (accessed July 18, 2018).

91. Ibid.

92. Ibid.

93. Ibid.

94. Ibid.

95. Ibid.

96. CNN Library, "2016 Presidential Campaign Hacking Fast Facts," CNN, May 16, 2018, http://www.cnn.com/2016/12/26/us/2016-presidential-campaign-hacking -fast-facts/index.html (accessed August 3, 2017).

97. Libicki, "China Developing Cyber Capabilities."

98. Nikolai Litovkin, "Russia Successfully Tests New Missile for Defense System Near Moscow," *Russia Beyond the Headlines,* June 23, 2016, https://www.rbth.com/ defence/2016/06/23/russia-successfully-tests-new-missile-for-defense-system-near -moscow_605711 (accessed August 2, 2017).

99. Ibid.

100. "Terminal High Altitude Area Defense," Lockheed Martin, 2018, http://www .lockheedmartin.com/us/products/thaad.html (accessed August 2, 2017).

101. Litovkin, "Russia Successfully Tests New Missile."

102. David Willman, "US Missile Defense System Is 'Simply Unable to Protect the Public,' Report Says," *Los Angeles Times,* July 14, 2016, http://www.latimes.com/projects/ la-na-missile-defense-failings (accessed August 2, 2017).

103. "Kalashnikov Gunmaker Develops Combat Module Based on Artificial Intelligence," TASS, July 5, 2016, http://tass.com/defense/954894 (accessed August 3, 2017).

104. Ibid.

105. "Russian Military to Deploy Security Bots at Missile Bases," Sputnik, March 13, 2014, https://sputniknews.com/russia/20140313188363867-Russian-Military-to-Deploy -Security-Bots-at-Missile-Bases (accessed August 3, 2017).

106. Tristan Greene, "Russia Is Developing AI Missiles to Dominate the New Arms Race," TNW, July 27, 2017, https://thenextweb.com/artificial-intelligence/2017/07/27/ russia-is-developing-ai-missiles-to-dominate-the-new-arms-race/#.tnw_NFwQAzWf (accessed August 3, 2017).

107. Dmitry Litovkin and Nikolai Litovkin, "Russia's Digital Doomsday Weapons: Robots Prepare for War," *Russia Beyond the Headlines*, May 31, 2017, https://www.rbth .com/defence/2017/05/31/russias-digital-weapons-robots-and-artificial-intelligence -prepare-for-wa_773677 (accessed August 3, 2017).

108. Rob Knake, "Russian Hackers Were Only Getting Started in the 2016 Election," *Fortune,* January 15, 2017, http://fortune.com/2017/01/15/russian-hackers-2016 -election-cyber-war (accessed August 3, 2017).

109. Ted Koppel, *Lights Out: A Cyberattack, A Nation Unprepared, Surviving the Aftermath* (New York: Broadway Books, 2015), p. 226.

110. *Business Blackout: The Insurance Implications of a Cyber Attack on the US Power Grid* (Emerging Risk Report; Cambridge, UK: Lloyd's of London and the University of Cambridge Centre for Risk Studies, 2015), https://www.lloyds.com/news-and-risk -insight/risk-reports/library/society-and-security/business-blackout (accessed April 13, 2018).

111. Michael Connell and Sarah Vogler, *Russia's Approach to Cyber Warfare* (Arlington, VA: Center for Naval Analysis, September 2016), http://www.dtic.mil/get-tr-doc/ pdf?AD=AD1019062 (accessed April 13, 2018).

CHAPTER 4: THE NEW REALITY

1. Frank Hoffman, "The Contemporary Spectrum of Conflict," in *2016 Index of US Military Strength* (Washington, DC: Heritage Foundation, 2016), http://index.heritage.org/military/2016/essays/contemporary-spectrum-of-conflict (accessed August 6, 2017).

2. Campaign to Stop Killer Robots, 2018, https://www.stopkillerrobots.org/about-us (accessed August 5, 2017).

3. Stuart Russell et al., "Autonomous Weapons: An Open Letter from AI & Robotics Researchers," Future of Life Institute, July 28, 2015, https://futureoflife.org/open-letter-autonomous-weapons (accessed August 5, 2017).

4. Richard Roth, "UN Security Council Imposes New Sanctions on North Korea," CNN, August 6, 2017, http://www.cnn.com/2017/08/05/asia/north-korea-un-sanctions/index.html (accessed August 6, 2017).

5. Warren Mass, "N. Korea Continues Missile Tests; US Moves 3rd Carrier Strike Force to Western Pacific," *New American*, May 29, 2017, https://www.thenewamerican.com/world-news/asia/item/26129-n-korea-continues-missile-tests-u-s-moves-3rd-carrier-strike-force-to-western-pacific (accessed August 6, 2017).

6. Franz-Stefan Gady, "Trump: 2 Nuclear Subs Operating in Korean Waters," *Diplomat*, May 25, 2017, http://thediplomat.com/2017/05/trump-2-nuclear-subs-operating-in-korean-waters (accessed August 6, 2017).

7. Jin Kai, "What THAAD Means for China's Korean Peninsula Strategy," *Diplomat*, July 27, 2017, http://thediplomat.com/2017/07/what-thaad-means-for-chinas-korean-peninsula-strategy (accessed August 6, 2017).

8. Hannah Beech, Yang Siqi, and Mark Thompson, "Inside the International Contest Over the Most Important Waterway in the World," *Time*, May 26, 2016, http://time.com/4348957/inside-the-international-contest-over-the-most-important-waterway-in-the-world (accessed August 6, 2017).

9. Will Nicol, "Showdown in the South China Sea: China's Artificial Islands Explained," Digital Trends, May 3, 2017, https://www.digitaltrends.com/cool-tech/chinas-artificial-islands-news-rumors (accessed August 6, 2017).

10. Volodymyr Valkov, "Expansionism: The Core of Russia's Foreign Policy," *New Eastern Europe*, August 12, 2014, http://neweasterneurope.eu/2014/08/12/expansionism-core-russias-foreign-policy/ (accessed August 6, 2017).

11. Adrian Bonenberger, "The War No One Notices in Ukraine," *New York Times*, June 20, 2017, https://www.nytimes.com/2017/06/20/opinion/ukraine-russia.html (accessed August 6, 2017).

12. "The Biological Weapons Convention," United Nations Office at Geneva, https://www.unog.ch/80256EE600585943/(httpPages)/04FBBDD6315AC720C12571 80004B1B2F?OpenDocument (accessed August 7, 2017).

13. "Convention on the Prohibition of the Development, Production, Stockpiling and Use of Chemical Weapons and on their Destruction," United Nations Treaty Collection, September 3, 1992, https://treaties.un.org/Pages/ViewDetails.aspx?src=TREATY&mtdsg_no=XXVI-3&chapter=26&lang=en (accessed August 7, 2017).

14. *United Nations Treaties and Principles on Outer Space* (New York: United Nations Office for Outer Space Affairs), ST/SPACE/61/Rev.1, http://www.unoosa.org/pdf/publications/ST_SPACE_061Rev01E.pdf (accessed August 7, 2017).

15. "Additional Protocol to the Convention on Prohibitions or Restrictions on the Use of Certain Conventional Weapons Which May Be Deemed to Be Excessively Injurious or to Have Indiscriminate Effects (Protocol IV, Entitled Protocol on Blinding Laser Weapons)," *United Nations Treaty Collection*, October 13, 1995, https://treaties.un.org/pages/ViewDetails.aspx?src=TREATY&mtdsg_no=XXVI-2-a&chapter=26&lang=en (accessed August 7, 2017).

16. Dan Drollette Jr., "Blinding Them with Science: Is Development of a Banned Laser Weapon Continuing?" *Bulletin of the Atomic Scientists*, September 14, 2014, http://thebulletin.org/blinding-them-science-development-banned-laser-weapon-continuing7598 (accessed August 7, 2017).

17. Michael R. Gordon, "US Says Russia Tested Cruise Missile, Violating Treaty," *New York Times*, July 28, 2014, https://www.nytimes.com/2014/07/29/world/europe/us-says-russia-tested-cruise-missile-in-violation-of-treaty.html?_r=0 (accessed August 5, 2017).

18. Michael R. Gordon, "Russia Deploys Missile, Violating Treaty and Challenging Trump," *New York Times*, February, 14, 2017, https://www.nytimes.com/2017/02/14/world/europe/russia-cruise-missile-arms-control-treaty.html (accessed August 5, 2017).

19. "Neurons & Synapses," The Human Memory, http://www.human-memory.net/brain_neurons.html (accessed August 8, 2017).

20. Susan Perry, "Glia: The Other Brain Cells," BrainFacts.org, September 15, 2010, http://www.brainfacts.org/Archives/2010/Glia-the-Other-Brain-Cells (accessed August 8, 2017).

21. "Neurons & Synapses."

22. Gideon Lewis-Kraus, "The Great AI Awakening," *New York Times*, December 14, 2016, https://www.nytimes.com/2016/12/14/magazine/the-great-ai-awakening.html?_r=0&mtrref=undefined (accessed August 9, 2017).

23. Marc Andreessen et al., "a16z Podcast: Software Programs the World," Andreessen Horowitz, July 10, 2016, https://a16z.com/2016/07/10/software-programs-the-world (accessed August 9, 2017).

24. Walden C. Rhines, "Moore's Law and the Future of Solid-State Electronics," *Scientific American*, April 12, 2016, https://blogs.scientificamerican.com/guest-blog/moore-s-law-and-the-future-of-solid-state-electronics/ (accessed July 18, 2018).

25. Editorial Team, "The Exponential Growth of Data," Inside BIGDATA, February 16, 2017, https://insidebigdata.com/2017/02/16/the-exponential-growth-of-data (accessed August 9, 2017).

26. "Ray Kurzweil," National Inventors Hall of Fame, 2016, http://www.invent.org/honor/inductees/inductee-detail/?IID=180 (accessed August 10, 2017).

27. "Chris F. Westbury," University of Alberta, https://sites.ualberta.ca/~chrisw (accessed August 10, 2017).

28. Chris F. Westbury, "On the Processing Speed of the Human Brain," *Chris F. Westbury* (blog), June 26, 2014, http://chrisfwestbury.blogspot.com/2014/06/on-processing-speed-of-human-brain.html (accessed August 10, 2017).

29. Vincent C. Müller and Nick Bostrom, "Future Progress in Artificial Intelligence: A Survey of Expert Opinion," in *Fundamental Issues of Artificial Intelligence*, ed. Vincent C. Müller (Berlin: Springer, 2014), https://nickbostrom.com/papers/survey.pdf (accessed August 10, 2017).

30. James Vincent, "Chinese Supercomputer Is the World's Fastest—and Without

Using US Chips," Verge, Jun 20, 2016, https://www.theverge.com/2016/6/20/
11975356/chinese-supercomputer-worlds-fastes-taihulight (accessed August 10, 2017).

31. Müller and Bostrom, "Future Progress in Artificial Intelligence."

32. Ibid.

33. Ray Kurzweil, *The Singularity Is Near: When Humans Transcend Biology* (New York: Viking, 2005), p. 136.

34. Ashton B. Carter, "Directive 3000.09: Autonomy in Weapon Systems" (Washington, DC: US Department of Defense Directive, November 21, 2012), p. 2, http://www.esd.whs.mil/Portals/54/Documents/DD/issuances/dodd/300009p.pdf (accessed August 11, 2017).

35. Matthew Rosenberg and John Markoff, "The Pentagon's 'Terminator Conundrum': Robots That Could Kill on Their Own," *New York Times*, October 25, 2016, https://www.nytimes.com/2016/10/26/us/pentagon-artificial-intelligence-terminator .html?_r=0 (accessed August 11, 2017).

36. Vasily Kashin, "Russia's S-400 to Help China Control Taiwan and Diaoyu Airspace—Expert," *Russia Beyond the Headlines*, February 20, 2017, https://www.rbth.com/opinion/2017/02/20/russia-s-400-china-taiwan-705823 (accessed August 11, 2017).

37. "S-400 Triumph Air Defense Missile System, Russia," Army Technology, http://www.army-technology.com/projects/s-400-triumph-air-defence-missile-system (accessed August 11, 2017).

38. Clare Wilson, "Maxed Out: How Many Gs Can You Pull?" *New Scientist*, April 14, 2010, https://www.newscientist.com/article/mg20627562-200-maxed-out-how -many-gs-can-you-pull (accessed August 11, 2017).

CHAPTER 5: DEVELOPING GENIUS WEAPONS

1. David Smalley, "The Future Is Now: Navy's Autonomous Swarmboats Can Overwhelm Adversaries," US Navy, Office of Naval Research, October 5, 2014, https://www.onr.navy.mil/Media-Center/Press-Releases/2014/autonomous-swarm-boat -unmanned-caracas.aspx (accessed August 13, 2017).

2. CNN Library, "USS Cole Bombing Fast Facts," CNN, June 2, 2017, http://www.cnn.com/2013/09/18/world/meast/uss-cole-bombing-fast-facts/index.html (accessed August 13, 2017).

3. Patrick Tucker, "The US Navy's Autonomous Swarm Boats Can Now Decide What to Attack," *Defense One*, December 14, 2016, http://www.defenseone.com/technology/2016/12/navys-autonomous-swarm-boats-can-now-decide-what-attack/ 133896 (accessed August 13, 2017).

4. Louis A. Del Monte, *Nanoweapons: A Growing Threat to Humanity* (Lincoln, NE: Potomac Books, 2017), p. 67.

5. Ibid., pp. 8–10.

6. Ibid., p. 46.

7. Ibid., p. 58.

8. Vincent C. Müller and Nick Bostrom, "Future Progress in Artificial Intelligence: A Survey of Expert Opinion," in *Fundamental Issues of Artificial Intelligence*, ed. Vincent C.

Müller (Berlin: Springer, 2014), https://nickbostrom.com/papers/survey.pdf (accessed August 10, 2017).

9. Ibid., p. 48.

10. Ibid.

11. Ibid.

12. Patrick Tucker, "The Military Wants Smarter Insect Spy Drones," *Defense One*, December 23, 2014, http://www.defenseone.com/technology/2014/12/military-wants -smarter-insect-spy-drones/101970 (accessed August 14, 2017).

13. Staff Reporter, "Botulinum Toxin Type H: The Deadliest Known Toxin with No Known Antidote Discovered," *Nature World News*, October 15, 2013, http://www .natureworldnews.com/articles/4442/20131015/botulinum-toxin-type-h-deadliest -known-antidote-discovered.htm (accessed August 14, 2017).

14. Del Monte, *Nanoweapons*, pp. 83–84.

15. Ibid., pp. 84–85.

16. Ibid., p. 183.

17. Leukemia Research Foundation, "Leukemia Research Foundation-Funded Researcher Utilizes Nanobots for Groundbreaking Leukemia Treatment," news release, January 6, 2016, http://www.allbloodcancers.org/index.cfm?fuseaction=news .details&ArticleId=74 (accessed August 16, 2017).

18. Brian Wang, "Pfizer Partnering with Ido Bachelet on DNA Nanorobots," *Next Big Future* (blog), May 15, 2015, https://www.nextbigfuture.com/2015/05/pfizer -partnering-with-ido-bachelet-on.html (accessed August 16, 2017).

19. Daniel Korn, "DNA Nanobots Will Target Cancer Cells in the First Human Trial Using a Terminally Ill Patient," *Plaid Zebra*, March 27, 2015, http://www.the plaidzebra.com/dna-nanobots-will-target-cancer-cells-in-the-first-human-trial-using-a -terminally-ill-patient (accessed August 16, 2017).

20. Tom Regan, "Nanobots Can Swim Your Bloodstream Faster by Doing the Front Crawl," *engadget*, July 25, 2017, https://www.engadget.com/2017/07/25/the-next-wave -of-nanobots-will-swim-front-crawl-in-your-blood (accessed August 16, 2017).

21. Del Monte, *Nanoweapons*, p. vii.

22. Eric Drexler, "Molecular Manufacturing Will Use Nanomachines to Build Large Products with Atomic Precision," E-drexler.com, http://e-drexler.com/p/04/03/ 0325molManufDef.html (accessed August 17, 2017).

23. Del Monte, *Nanoweapons*, chapter 2.

24. Eric Drexler, "'There's Plenty of Room at the Bottom' (Richard Feynman, Pasadena, 29 December 1959)," *Metamodern*, December 29, 2009, http://metamodern .com/2009/12/29/theres-plenty-of-room-at-the-bottom"-feynman-1959/ (accessed August 17, 2017).

25. Andrea Thompson, "Nanotech Produces Plastic as Strong as Steel," Innovation, NBC News, October 12, 2007, http://www.nbcnews.com/id/21268376/ns/technology _and_science-innovation/t/nanotech-produces-plastic-strong-steel/#.WZXxJ1WGPIU (accessed August 17, 2017).

26. Del Monte, *Nanoweapons*, p. 17.

27. G. I. Yakovlev et al., "Modification of Cement Matrix Using Carbon Nanotube Dispersions and Nanosilica," *Procedia Engineering* 172 (2017): 1261–69, http://www .sciencedirect.com/science/article/pii/S1877705817306549# (accessed August 17, 2017).

28. Del Monte, *Nanoweapons*, pp. 55–56.

29. Neil Gershenfeld and Isaac L. Chuang, "Quantum Computing with Molecules," *Scientific American*, June 1998, http://cba.mit.edu/docs/papers/98.06.sciqc.pdf (accessed August 18, 2017).

30. Sophia Chen, "Chinese Satellite Relays a Quantum Signal between Cities," *Wired*, June 15, 2017, https://www.wired.com/story/chinese-satellite-relays-a-quantum-signal-between-cities (accessed August 18, 2017).

31. Richard Haughton, "Quantum Teleportation Is Even Weirder Than You Think," *Nature*, July 20, 2017, https://www.nature.com/news/quantum-teleportation-is-even-weirder-than-you-think-1.22321 (accessed July 14, 2018).

32. Charles H. Bennett, "Notes on Landauer's Principle, Reversible Computation and Maxwell's Demon," *Studies in History and Philosophy of Modern Physics* 34, no. 3 (September 2003): 501–510, https://arxiv.org/pdf/physics/0210005.pdf (accessed July 15, 2018).

CHAPTER 6: CONTROLLING AUTONOMOUS WEAPONS

1. H+Pedia, s.v. "Artificial Life," last edited December 31, 2015, https://hpluspedia.org/wiki/Artificial_Life (accessed August 21, 2017).

2. Thomas Ray, "An Approach to the Synthesis of Life," *Artificial Life II, Santa Fe Institute Studies in the Sciences of Complexity*, ed. Christopher G. Langton et al., vol. 11 (Boston: Addison-Wesley, 1991), pp. 371–408, http://life.ou.edu/pubs/alife2/tierra.tex (accessed August 21, 2017).

3. Arthur C. Clarke, *2010: Odyssey Two* (New York: Del Rey, 1984), pp. 255–61.

4. Joanne Pransky, "The Essential Interview: Gianmarco Veruggio, Telerobotics and 'Roboethics' Pioneer," *Robotics Business Review*, February 1, 2017, https://www.roboticsbusinessreview.com/research/essential-interview-gianmarco-veruggio-telerobotics-roboethics-pioneer (accessed August 21, 2017).

5. Kristina Grifantini, "Robots 'Evolve' the Ability to Deceive," *MIT Technology Review*, August 18, 2009, https://www.technologyreview.com/s/414934/robots-evolve-the-ability-to-deceive (accessed August 23, 2017).

6. Ibid.

7. *Brain Waves Module 3: Neuroscience, Conflict, and Security* (London: Royal Society, February 2012), https://royalsociety.org/~/media/Royal_Society_Content/policy/projects/brain-waves/2012-02-06-BW3.pdf (accessed August 23, 2017).

8. V. P. Clark et al., "TDCS Guided Using fMRI Significantly Accelerates Learning to Identify Concealed Objects," *Neuroimage* 59, no. 1 (January 2, 2012): 117–28, https://www.ncbi.nlm.nih.gov/pubmed/21094258 (accessed August 23, 2017).

9. *Brain Waves Module 3*.

10. Johns Hopkins Medicine, "Mind-Controlled Prosthetic Arm Moves Individual 'Fingers,'" news release, February 15, 2016, https://www.hopkinsmedicine.org/news/media/releases/mind_controlled_prosthetic_arm_moves_individual_fingers (accessed August 24, 2017).

11. Rachel Metz, "Mind-Controlled VR Game Really Works," *MIT Technology*

Review, August 9, 2017, https://www.technologyreview.com/s/608574/mind-controlled
-vr-game-really-works (accessed August 24, 2017).

   12. José Delgado and Hannibal Hamlin, "Surface and Depth Electrography of
the Frontal Lobes in Conscious Patients," *Electroencephalography and Clinical Neurophysiology*
8, no. 3 (August 1956): 371–84, http://www.sciencedirect.com/science/article/
pii/0013469456900037 (accessed August 24, 2017).

   13. Max O. Krucoff et al., "Enhancing Nervous System Recovery through Neuro-
biologics, Neural Interface Training, and Neurorehabilitation," *Frontiers in Neuroscience*,
10, no. 584 (December 27, 2016), https://www.ncbi.nlm.nih.gov/pmc/articles/
PMC5186786 (accessed August 24, 2017).

   14. C. Hammond et al., "Latest View on the Mechanism of Action of Deep Brain
Stimulation," *Movement Disorders* 23, no. 15 (2008): 2111–21.

   15. Vincent C. Müller and Nick Bostrom, "Future Progress in Artificial Intelligence:
A Survey of Expert Opinion," in *Fundamental Issues of Artificial Intelligence*, ed. Vincent C.
Müller (Berlin: Springer, 2014), https://nickbostrom.com/papers/survey.pdf (accessed
August 10, 2017).

   16. Alice Park, "There's No Known Limit to How Long Humans Can Live,
Scientists Say," *Time*, June 28, 2017, http://time.com/4835763/how-long-can-humans
-live (accessed July 20, 2018).

   17. Elizabeth Arias, "United States Life Tables, 2003," *National Vital Statistics Reports*
54, no. 14 (April 19, 2006; revised March 28, 2007), https://www.cdc.gov/nchs/data/
nvsr/nvsr54/nvsr54_14.pdf (accessed July 20, 2018).

   18. Fiona Macdonald, "A Robot Has Just Passed a Classic Self-Awareness Test for
the First Time," *Science Alert*, July 17, 2015, https://www.sciencealert.com/a-robot-has
-just-passed-a-classic-self-awareness-test-for-the-first-time (accessed August 24, 2017).

   19. Ibid.

CHAPTER 7: THE ETHICAL DILEMMAS

   1. Campaign to Stop Killer Robots, 2018, https://www.stopkillerrobots.org
(accessed August 29, 2017).

   2. "The Problem," Campaign to Stop Killer Robots, http://www.stopkillerrobots
.org/the-problem (accessed August 29, 2017).

   3. "Concern from the United Nations," Campaign to Stop Killer Robots, https://
www.stopkillerrobots.org/2017/07/unitednations (accessed August 29, 2017).

   4. Ibid.

   5. "Fully Autonomous Weapons," Reaching Critical Will, http://www.reaching
criticalwill.org/resources/fact-sheets/critical-issues/7972-fully-autonomous-weapons
(accessed August 28, 2017).

   6. Women's International League for Peace and Freedom, https://wilpf.org
(accessed July 23, 2018).

   7. Ashton B. Carter, "Directive 3000.09: Autonomy in Weapon Systems" (Washington, DC: US Department of Defense Directive), November 21, 2012, pp. 13–14, http://
www.esd.whs.mil/Portals/54/Documents/DD/issuances/dodd/300009p.pdf (accessed
July 26, 2017).

8. Vincent C. Müller and Nick Bostrom, "Future Progress in Artificial Intelligence: A Survey of Expert Opinion," in *Fundamental Issues of Artificial Intelligence*, ed. Vincent C. Müller (Berlin: Springer, 2014), https://nickbostrom.com/papers/survey.pdf (accessed July 24, 2018).

9. Seth Thornhill, "Future Autonomous Robotic Systems in the Pacific Theater" (master's thesis, Joint Advanced Warfighting School, May 6, 2015), p. 20, http://www.dtic.mil/dtic/tr/fulltext/u2/a624818.pdf (accessed August 29, 2017).

10. Patrick Tucker, "Russian Weapons Maker to Build AI-Directed Guns," *Defense One*, July 14, 2017, https://www.defenseone.com/technology/2017/07/russian-weapons-maker-build-ai-guns/139452 (accessed July 25, 2018).

11. *The United States Strategic Bombing Survey, Summary Report: European War* ([Washington, DC: US Government Printing Office,] September 30, 1945), p. 5, http://www.anesi.com/ussbs02.htm (accessed August 29, 2017).

12. Malcolm W. Browne, "Invention That Shaped the Gulf War: The Laser-Guided Bomb," *New York Times*, February 26, 1991, http://www.nytimes.com/1991/02/26/science/invention-that-shaped-the-gulf-war-the-laser-guided-bomb.html (accessed August 29, 2017).

13. "Reagan National Defense Forum Keynote as Delivered by Secretary of Defense Chuck Hagel," Ronald Reagan Presidential Library, November 15, 2014, https://www.defense.gov/News/Speeches/Speech-View/Article/606635 (accessed August 30, 2017).

14. Ibid.

15. John Markoff and Matthew Rosenberg, "China's Intelligent Weaponry Gets Smarter," *New York Times*, February 3, 2017, https://www.nytimes.com/2017/02/03/technology/artificial-intelligence-china-united-states.html (accessed July 24, 2017).

16. Carter, "Directive 3000.09," p. 2.

17. April Glaser, "The UN Has Decided to Tackle the Issue of Killer Robots in 2017," Recode, December 16, 2016, https://www.recode.net/2016/12/16/13988458/un-killer-robots-elon-musk-wozniak-hawking-ban (accessed August 30, 2017).

18. Harold C. Hutchison, "Russia Says It Will Ignore Any UN Ban of Killer Robots," *Business Insider*, November 30, 2017, https://www.businessinsider.com/russia-will-ignore-un-killer-robot-ban-2017-11 (accessed July 25, 2018).

19. Tucker Davey, "Lethal Autonomous Weapons: An Update from the United Nations," Future of Life Institute, April 30, 2018, https://futureoflife.org/2018/04/30/lethal-autonomous-weapons-an-update-from-the-united-nations/?cn-reloaded=1 (accessed July 25, 2018).

20. Steven Groves, "The US Should Oppose the UN's Attempt to Ban Autonomous Weapons," Heritage Foundation, March 5, 2015, https://www.heritage.org/defense/report/the-us-should-oppose-the-uns-attempt-ban-autonomous-weapons (accessed August 30, 2017).

21. Office of the Historian, "Milestones: 1969–1976: Strategic Arms Limitations Talks/Treaty (SALT) I and II," US Department of State, https://history.state.gov/milestones/1969-1976/salt (accessed August 31, 2017).

22. "The Treaty between the United States of America and the Union of Soviet Socialist Republics on the Reduction and Limitation of Strategic Offensive Arms (START)," US Department of State, July 31, 1991, October 2001 ed., last revised May 2002, https://www.state.gov/t/avc/trty/146007.htm (accessed August 31, 2017).

23. "New START," US Department of State, February 5, 2011, https://www.state .gov/t/avc/newstart (accessed August 31, 2017).

24. Groves, "US Should Oppose the UN's Attempt."

25. Ibid.

26. Heather M. Roff, "What Do People Around the World Think about Killer Robots?" *Slate*, February 8, 2017, http://www.slate.com/articles/technology/future _tense/2017/02/what_do_people_around_the_world_think_about_killer_robots.html (accessed August 31, 2017).

27. Ibid.

28. John Lewis, "The Case for Regulating Fully Autonomous Weapons," *Yale Law Journal* 124, no. 4 (January–February 2015), https://www.yalelawjournal.org/comment/ the-case-for-regulating-fully-autonomous-weapons (accessed August 31, 2017).

29. Protocol Additional to the Geneva Conventions of August 12, 1949, and relating to the Protection of Victims of International Armed Conflicts (Protocol I). June 8, 1977, art. 51(5)(b), https://www.icrc.org/eng/assets/files/other/icrc_002_0321.pdf (accessed July 20, 2018).

30. Lewis, "Case for Regulating."

31. Carl Hoffman, "China's Space Threat: How Missiles Could Target US Satellites," *Popular Mechanics*, December 17, 2009, http://www.popularmechanics.com/ space/satellites/a1782/4218443 (September 1, 2017).

32. Paul Bedard, "Congress Warned North Korean EMP Attack Would Kill '90% of All Americans,'" *Washington Examiner*, October 12, 2017, https://www.washington examiner.com/congress-warned-north-korean-emp-attack-would-kill-90-of-all-americans (accessed July 25, 2018).

33. Ferris Jabr, "Know Your Neurons: What Is the Ratio of Glial to Neurons in the Brain?" *Scientific American*, June 13, 2012, https://blogs.scientificamerican.com/ brainwaves/know-your-neurons-what-is-the-ratio-of-glia-to-neurons-in-the-brain/ (accessed September 2, 2017).

34. Müller and Bostrom, "Future Progress in Artificial Intelligence."

CHAPTER 8: WAR ON AUTOPILOT

1. Erik Sofge, "Tale of the Teletank: The Brief Rise and Long Fall of Russia's Military Robots," *Popular Science*, March 7, 2014, http://www.popsci.com/blog-network/ zero-moment/tale-teletank-brief-rise-and-long-fall-russia%E2%80%99s-military-robots (accessed September 8, 2017).

2. Ibid.

3. Dave Majumdar, "Russia's Lethal New Robotic Tanks Are Going Global," *National Interest*, February 8, 2016, http://nationalinterest.org/blog/russias-lethal-new -robotic-tanks-are-going-global-15143 (accessed September 8, 2017).

4. Ibid.

5. Aric Jenkins, "The USS Gerald Ford Is the Most Advanced Aircraft Carrier in the World," *Fortune*, July 22, 2017, http://fortune.com/2017/07/22/uss-gerald-ford -commissioning (accessed September 8, 2017).

6. "Nimitz Class Aircraft Carrier, United States of America," Naval Technology, http://www.naval-technology.com/projects/nimitz (accessed September 8, 2017).

7. "DDG-51 Arleigh Burke-Class," GlobalSecurity.org, https://www.globalsecurity .org/military/systems/ship/ddg-51.htm (accessed September 8, 2017).

8. "Prepared to Defend," *All Hands*, http://www.navy.mil/ah_online/zumwalt (accessed September 8, 2017).

9. Advisory Service on International Humanitarian Law, "What Is International Humanitarian Law?" (Geneva: International Committee of the Red Cross, July 2004), https://www.icrc.org/eng/assets/files/other/what_is_ihl.pdf (accessed September 8, 2017).

10. John Naisbitt, *Megatrends: Ten New Directions Transforming Our Lives* (New York: Warner Books, October 27, 1982).

11. Erwin Chemerinsky, *Criminal Procedure: Adjudication*, 2nd ed. (New York: Aspen Publishers, July 29, 2013), p. 221.

12. Office of Public Health Preparedness and Response, "Possible Health Effects of Radiation Exposure and Contamination," Center for Disease Control and Prevention, last updated October 10, 2014, https://emergency.cdc.gov/radiation/healtheffects.asp (accessed September 9, 2017).

13. Zoe T. Richards et al., "Bikini Atoll Coral Biodiversity Resilience Five Decades after Nuclear Testing," *Marine Pollution Bulletin* 56, no. 3 (March 2008): 503–15.

14. *Nuclear Weapons and International Humanitarian Law*, Information Note no. 4 (Geneva, Switzerland: International Committee of the Red Cross, March 3, 2013), https://www.icrc.org/eng/resources/documents/legal-fact-sheet/03-19-nuclear -weapons-ihl-4-4132.htm (accessed September 9, 2017).

15. George Sylvester Viereck, *Saturday Evening Post*, October, 26, 1929, p. 17.

16. "Why Do We Age and Is There Anything We Can Do about It?" The Tech, https://genetics.thetech.org/original_news/news10 (accessed September 10, 2017).

17. Ibid.

18. Anthony Atala, "Growing New Organs," TEDMED 2009, October 2009, Ted, 17:45, https://www.ted.com/talks/anthony_atala_growing_organs_engineering _tissue?language=en (accessed September 10, 2017).

19. "Nearby Super-Earth Likely a Diamond Planet," *Yale News*, October 11, 2012, https://news.yale.edu/2012/10/11/nearby-super-earth-likely-diamond-planet (accessed September 11, 2017).

20. *United Nations Treaties and Principles on Outer Space* (New York: United Nations Office For Outer Space Affairs), ST/SPACE/61/Rev.1, http://www.unoosa.org/pdf/ publications/ST_SPACE_061Rev01E.pdf (accessed September 11, 2017).

21. Wikipedia, s.v. "Militarization of Space," last edited June 20, 2018, https:// en.wikipedia.org/wiki/Militarisation_of_space#Outer_Space_Treaty (accessed June 26, 2018).

22. "Proposed Prevention of an Arms Race in Space (PAROS) Treaty," Nuclear Threat Initiative (NTI), last updated September 29, 2017, http://www.nti.org/learn/ treaties-and-regimes/proposed-prevention-arms-race-space-paros-treaty/ (accessed July 30, 2018).

23. Brendan Nicholson, "World Fury at Satellite Destruction," *The Age* (Melbourne, Australia), January 20, 2007, https://www.theage.com.au/national/world-fury-at-satellite -destruction-20070120-ge416d.html (accessed July 26, 2018).

24. Harsh Vasani, "How China Is Weaponizing Outer Space," *The Diplomat*, January 19, 2017, http://thediplomat.com/2017/01/how-china-is-weaponizing-outer-space (accessed September 12, 2017).

25. Ibid.

26. Ibid.

27. "How Do Satellites Survive Hot and Cold Orbit Environments?" *Astrome* (blog), July 22, 2015, http://www.astrome.co/blogs/how-do-satellites-survive-hot-and-cold-orbit -environments (accessed September 12, 2017).

28. Karl Tate, "Space Radiation Threat to Astronauts Explained," Space.com, May 30, 2013, https://www.space.com/21353-space-radiation-mars-mission-threat.html (accessed September 12, 2017).

29. "Understanding Space Radiation," *NASA Facts*, October 2002, https://spaceflight .nasa.gov/spacenews/factsheets/pdfs/radiation.pdf (accessed September 12, 2017).

30. Tate, "Space Radiation Threat."

CHAPTER 9: WHO IS THE ENEMY?

1. Vincent C. Müller and Nick Bostrom, "Future Progress in Artificial Intelligence: A Survey of Expert Opinion," in *Fundamental Issues of Artificial Intelligence*, ed. Vincent C. Müller (Berlin: Springer, 2014), https://nickbostrom.com/papers/survey.pdf (accessed August 10, 2017).

2. K. Eric Drexler, *Engines of Creation: The Coming Era of Nanotechnology* (New York: Anchor Books, 1987), pp. 53–63.

3. Inbal Wiesel-Kapah et al., "Rule-Based Programming of Molecular Robot Swarms for Biomedical Applications," *Proceedings of the Twenty-Fifth International Joint Conference on Artificial Intelligence*, July 2016, https://www.ijcai.org/Proceedings/16/ Papers/495.pdf (accessed September 14, 2017).

4. Subbarao Kambhampati, ed., *Proceedings of the Twenty-Fifth International Joint Conference on Artificial Intelligence*, International Joint Conferences on Artificial Intelligence Organization, Palo Alto, CA, July 9–15, 2016, https://www.ijcai.org/proceedings/2016 (accessed September 14, 2017).

5. Wiesel-Kapah et al, "Rule-Based Programming."

6. Peter Rüegg, "Nanoscale Assembly Line," Eidgenössische Technische Hochschule Zürich, August 26, 2014, https://www.ethz.ch/en/news-and-events/eth -news/news/2014/08/Nanoscale-assembly-line.html (accessed September 14, 2017).

7. "IBM Research: Major Nanoscale Breakthroughs," IBM, http://www-03.ibm .com/press/attachments/28488.pdf (accessed September 14, 2017).

8. Larisa Brown, "Now You Can Be Bugged Anywhere: Military Unveils Insect-Sized Spy Drone with Dragonfly-Like Wings," *Daily Mail*, August 11, 2016, http://www .dailymail.co.uk/sciencetech/article-3734945/Now-bugged-Military-unveils-insect-sized -spy-drone-dragonfly-like-wings.html (accessed September 14, 2017).

9. Carl Von Clausewitz, *On War*, ed. and trans. Michael Howard and Peter Paret (Princeton, NJ: Princeton University Press, 1976), https://docentes.fd.unl.pt/docentes _docs/ma/FPG_MA_31565.pdf (accessed September 15, 2017).

10. Ibid., p. 101.

11. Ibid., p. 108.

12. Ibid., p. 120.

13. Lonsdale Hale, *The Fog of War* (Charing Cross: Edward Stanford, March 24, 1896).

14. Wikipedia, s.v. "*The Fog of War*" (2003 American documentary film), last edited June 9, 2018, https://en.wikipedia.org/wiki/The_Fog_of_War (accessed June 26, 2018).

15. *Wikipedia*, s.v. "Operation Fortitude," last edited June 18, 2018, https://en.wikipedia.org/wiki/Operation_Fortitude (accessed June 26, 2018).

16. Blake Stilwell, "The Army Built a Fake Base to Fool Saddam Hussein, and It Worked," *Business Insider*, October 9, 2015, https://www.businessinsider.com/the-army-built-a-fake-base-to-fool-saddam-hussein-and-it-worked-2015-10 (accessed July 29, 2018).

17. Diane Maye, "History's Last Left Hook?" *Medium*, December 14, 2015, https://medium.com/@DianeLeighMaye/history-s-last-left-hook-4711b1768cdd (accessed July 30, 2018).

18. Agence France Presse, "Putin Describes Secret Operation to Seize Crimea," Yahoo! News, March 8, 2015, https://www.yahoo.com/news/putin-describes-secret-operation-seize-crimea-212858356.html (accessed September 15, 2017).

19. Stephen A. Cook, "The Complexity of Theorem-Proving Procedures," Proceeding STOC '71 Proceedings of the Third Annual ACM Symposium on Theory of Computing, Shaker Heights, OH, May 3–5, 1971, pp. 151–58, http://dl.acm.org/citation.cfm?coll=GUIDE&dl=GUIDE&id=805047 (accessed September 17, 2017).

20. Adam Morton, *Emotion and Imagination* (Cambridge, UK: Polity, 2013), p. 3.

21. Müller and Bostrom, "Future Progress in Artificial Intelligence."

22. Aristotle, *Rhetoric Book II*, written 350 BCE, trans. W. Rhys Roberts, http://classics.mit.edu/Aristotle/rhetoric.2.ii.html (accessed September 18, 2017).

23. Robert Plutchik, *Circumplex Models of Personality and Emotions*, 1st ed. (Washington, DC: American Psychological Association, 1997), pp. 17–45.

24. Timothy Williamson, "Reclaiming the Imagination," *The Stone* (blog), *New York Times*, August 15, 2010, https://opinionator.blogs.nytimes.com/2010/08/15/reclaiming-the-imagination/?mcubz=0&_r=0 (accessed September 19, 2017).

25. Ibid.

26. John Montgomery, "Emotions, Survival, and Disconnection," *Embodied Mind* (blog), *Psychology Today*, September 30, 2012, https://www.psychologytoday.com/blog/the-embodied-mind/201209/emotions-survival-and-disconnection (accessed September 19, 2017).

27. Ibid.

28. "History: The Organization of the National Research Council," National Academies of Sciences, http://www.nasonline.org/about-nas/history/archives/milestones-in-NAS-history/organization-of-the-nrc.html (accessed July 29, 2018).

29. Jim Blascovich, Christine R Harte, and the National Research Council, *Human Behavior in Military Contexts* (Washington, DC: National Academies Press, February 3, 2008), pp. 55–63.

30. Karl Smallwood, "The Top 10 Most Inspiring Self-Sacrifices," Listverse, January 15, 2013, http://listverse.com/2013/01/15/the-top-10-most-inspiring-self-sacrifices (accessed September 20, 2017).

31. John 15:13 (King James Version).

32. "Fall of the Soviet Union," History, http://www.history.com/topics/cold-war/fall-of-soviet-union (accessed September 20, 2017).

33. *Wikipedia*, s.v. "Cold War," last edited June 26, 2018, https://en.wikipedia.org/wiki/Cold_War (accessed September 20, 2017).

34. Scott Sumner, "The Soviet Union: Military Spending," *Nintil* (blog), May 31, 2016, https://nintil.com/2016/05/31/the-soviet-union-military-spending (accessed September 20, 2017).

35. Ibid.

36. "Fall of the Soviet Union."

37. Albert Einstein, "To The General Assembly of the United Nations," open letter, United Nations World New York, October 1947, October 1947, http://neutrino.aquaphoenix.com/un-esa/ws1997-letter-einstein.html (accessed September 20, 2017).

38. Ronald Reagan, "Address to the 42d Session of the United Nations General Assembly in New York, New York," September 21, 1987, Ronald Reagan Presidential Library and Museum, https://www.reaganlibrary.gov/research/speeches/092187b (accessed July 30, 2018).

39. "Mutual Assured Destruction," Nuclear Age Peace Foundation, 2018, http://www.nuclearfiles.org/menu/key-issues/nuclear-weapons/history/cold-war/strategy/strategy-mutual-assured-destruction.htm (accessed September 21, 2017).

40. Alfred Werner, *Liberal Judaism* 16 (April–May 1949), Einstein Archive 30-1104, as sourced in Albert Einstein, *The New Quotable Einstein*, ed. Alice Calaprice (Princeton, NJ: Princeton University Press, 2005), p. 173.

CHAPTER 10: HUMANITY VERSUS MACHINE

1. Sun Tzu, *The Art Of War*, trans. Lionel Giles (Norwalk, CT: Puppet Press, 1910).

2. John 14:3 (King James Version).

3. "The Biological Weapons Convention," United Nations Office for Disarmament Affairs, New York, https://www.un.org/disarmament/wmd/bio (accessed September 22, 2017).

4. *Universal Declaration of Human Rights* (New York: United Nations, December 10, 1948), http://www.un.org/en/universal-declaration-human-rights/index.html (accessed September 23, 2017).

5. Isaac Asimov, "Runaround," in *I, Robot* (New York: Doubleday, 1950), p. 40.

6. Wikipedia, s.v. "Three Laws of Robotics," last edited May 7, 2018, https://en.wikipedia.org/wiki/Three_Laws_of_Robotics (accessed June 26, 2018).

7. James Gunn, *Isaac Asimov: The Foundations of Science Fiction, Revised Edition* (Lanham, MD: Scarecrow Press, 2005), p. 48.

EPILOGUE

1. Maureen Dowd, "Elon Musk's Billion-Dollar Crusade to Stop The AI Apocalypse," *Vanity Fair*, April 2017, https://www.vanityfair.com/news/2017/03/elon -musk-billion-dollar-crusade-to-stop-ai-space-x (accessed September 24, 2017).

2. Rory Cellan-Jones, "Stephen Hawking Warns Artificial Intelligence Could End Mankind," BBC, December 2, 2014, http://www.bbc.com/news/technology-30290540 (accessed September 24, 2017).

3. James Barrat, *Our Final Invention: Artificial Intelligence and the End of the Human Era* (New York: Thomas Dunne Books, 2013).

4. Nick Bostrom, *Superintelligence: Paths, Dangers, Strategies* (New York: Oxford University Press, 2014).

5. Louis A. Del Monte, *The Artificial Intelligence Revolution: Will Artificial Intelligence Serve Us or Replace Us?* (self-pub., April 17, 2014).

6. James Barrat, *Our Final Invention: Artificial Intelligence and the End of the Human Era* (New York: St. Martin's, 2015).

7. "June 2018," Top500, https://www.top500.org/lists/2018/06 (accessed July 31, 2018).

8. J. K. Rowling, "The Fringe Benefits of Failure, and the Importance of Imagination," *Harvard Gazette*, June 5, 2008, https://news.harvard.edu/gazette/story/ 2008/06/text-of-j-k-rowling-speech (accessed September 25, 2017).

9. "Thomas A. Edison Quotes," Goodreads, https://www.goodreads.com/author/ quotes/3091287.Thomas_A_Edison (accessed September 25, 2017).

INDEX

humans choosing to increase via brain implants (*see* SAIH [strong artificially intelligent human])
judging a computer's, 26, 248 (*see also* superintelligences)
    no machines currently exceed human-level general intelligence, 35
people recognizing device's function but not its intelligence, 34
See also strong artificial intelligence (strong AI)
intelligence, surveillance, and reconnaissance platforms (ISR), 65, 74, 81, 204
intelligent agent, 265
intelligent personal assistants (IPA), 52, 59
Intelligent Personal Devices (IPD), 52–53
Intelligent Tutor Systems (ITS), 58
interconnectivity of AI, 117–18
Intermediate-Range Nuclear Forces Treaty (INF), 101
International Committee of the Red Cross, 192
international humanitarian law (IHL), 207
    applying to AI and autonomous weapons, 12, 160, 164, 173, 174, 180, 189
    and humans on-the-loop, 187, 190, 251
    prohibiting nuclear weapons, 191, 192
international human rights law (IHRL), 160
International Joint Conference on Artificial Intelligence (24th conference), 95
    letter on threat of an AI arms race, 160–61, 259–60
international space laws, 201–205
International Space Station (ISS), 204–205
"Internet of Things" (IoT), 49–50, 106, 109, 117–18
in-the-loop: human controlled weapons, 70–71, 82, 110, 139, 147, 163, 187
IoT ("Internet of Things), 49–50, 106, 109, 117–18
IPA (intelligent personal assistants), 52, 59
IPD (Intelligent Personal Devices), 52–53

iPhone, 32, 50, 52–53
IPsoft, 40
IPSOS (polling firm), 173–74
Iran, 90, 208
iRobot, 57
irony of controlling genius weapons after singularity, 154–55
Islamic State (ISIS), use of drones, 67
ISR (intelligence, surveillance, and reconnaissance platforms), 65, 74, 81, 204
Israel Aerospace Industries, Ltd., 173
ISS (International Space Station), 204–205
ITS (Intelligent Tutor System), 58

JASSM (Joint Air-to-Surface Standoff Missile), 83
Jennings, Ken, 29
Jeopardy (TV quiz show), 29, 227
JFHQ-C (Commander, Joint Force Headquarters), 258
Jiuquan Satellite Launch Center, 133–34
Jobs, Steve, 52
Johns Hopkins on mind-controlled artificial arm, 146
Joint Air-to-Surface Standoff Missile (JASSM), 83

Kalashnikov Group, 89–90
Kasparov, Garry, 25, 29
killer bees, Navy's swarmboats acting as, 120
killer robots, 12–13, 68, 95, 161, 162
Koppel, Ted, 90
Korean War, 237
Korn, Daniel, 128
Kotov, Nicholas, 130
Kurzweil, Ray, 32, 107, 108, 254

Laboratory of Intelligent Systems (Swiss Federal Institute of Technology). *See* Lausanne experiment
Landauer, Rolf, 135
landmines, similarities to autonomous weapons, 174
lasers, banning use of to blind enemy, 98, 100, 171, 189, 260